FOCUS ON

MIDDLE SCHOOL

Grade

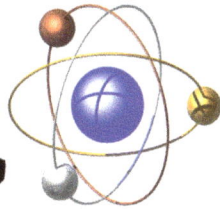

PHYSICS

3rd Edition

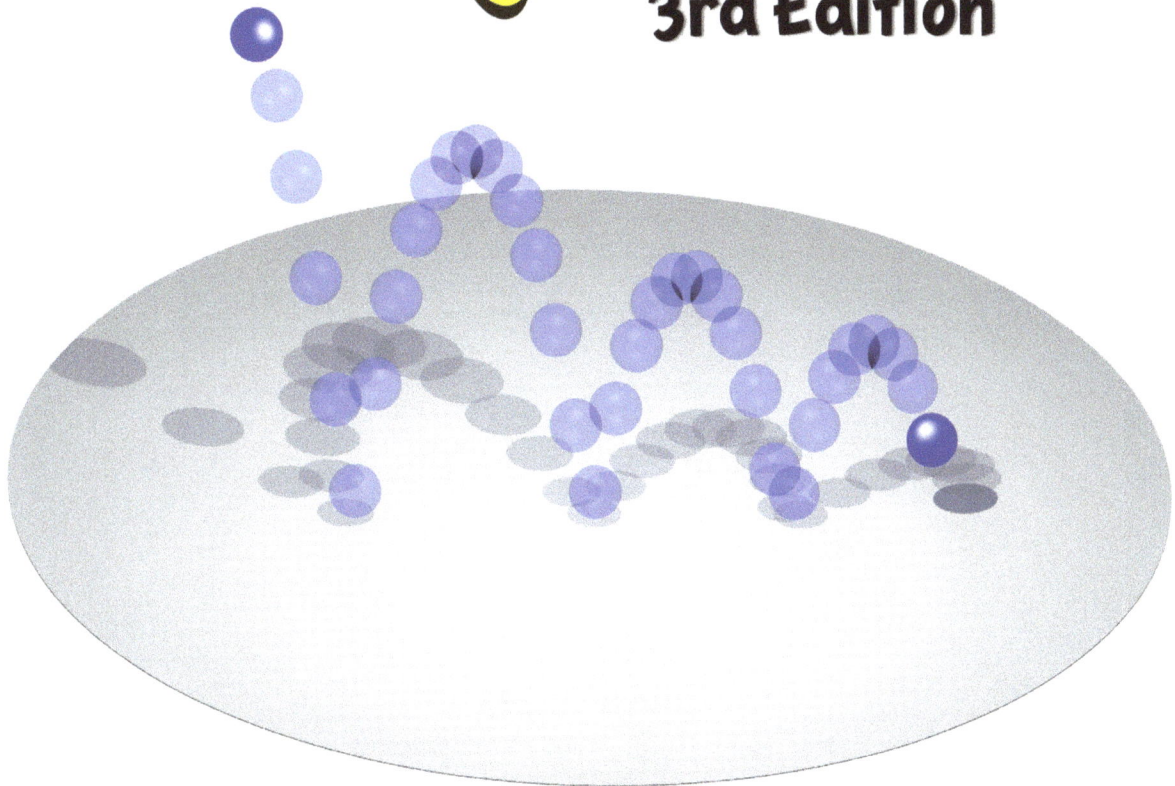

Rebecca W. Keller, PhD

Real Science-4-Kids

Illustrations: Janet Moneymaker

Focus On Middle School Physics Student Textbook—3rd Edition (softcover)
ISBN 978-1-941181-72-0

Published by Gravitas Publications Inc.
www.gravitaspublications.com
www.realscience4kids.com

GRAVITAS
PUBLICATIONS

Contents

Chapter 1 What Is Physics?

1.1 Introduction

Have you ever wondered what makes a feather float but a boulder fall, or why a bird can fly but a whale can't fly? Have you ever noticed that when your mom quickly puts on the brakes, the car stops, but your ice cream ends up on the dashboard? Have you ever wondered why, when you slide your stocking feet on the carpet, you can "shock" your dad?

All of these observations, and others like them, begin the inquiry into the field of science called physics. The name physics comes from the Greek word *physika*, which means "physical or natural." Physics investigates the most basic laws that govern the physical or natural world.

1.2 The Basic Laws of Physics

What is a basic law of physics? Are the laws of physics like the laws that tell us not to speed or not to steal? No. In fact, physical laws are statements that tell us about how the physical world works. Using these laws, we can understand why baseballs go up and then come down, why airplanes can fly, why rockets can land on the Moon, and why we see rainbows after it rains.

Physical laws are never broken, unlike laws that tell us not to speed or not to steal. For example, Newton's law of gravity tells us why we stay firmly on the surface of the Earth and do not sometimes just fly off. People have always known that the world behaves in regular and reliable ways. For example, people have observed for centuries that the Sun always rises and sets, that water always flows downhill, or that if it is cold enough, water will turn into ice. The laws of physics are statements about these regular and reliable observations.

We know that objects such as baseballs, airplanes, and people consistently obey the laws of physics and don't suddenly break one or two. It would be hard to play baseball if every once in a while the ball hit by the batter landed on the Moon!

1.3 How We Get Laws

How do we know what these laws are, and how did we discover them? Did the Earth come with a big instruction book that spelled out all of the laws? Not exactly. People had to figure them out on their own. Scientists use scientific investigation to discover how the world works.

One early scientist who used scientific investigation and helped develop the scientific method was Galileo Galilei. Galileo was an Italian astronomer born in Pisa, Italy in 1564. He showed how two lead balls fall at the same rate even if one is larger than the other. He performed a famous experiment where he is said to have dropped two cannon balls off the Leaning Tower of Pisa. He found that, even though the two cannon balls were different weights, they landed on the ground at exactly the same time!

People had trouble believing the results of Galileo's experiments, and it wasn't until Isaac Newton showed mathematically why this was true that it was finally accepted. Isaac Newton is considered to be one of the greatest scientists of all time. He is also considered to be the founder of physics as we know it today.

SIR ISAAC NEWTON
1643-1727 CE

Sir Isaac Newton was born on January 4, 1643 in Woolsthorpe, Lincolnshire, England. When Newton was 18 years old, he went to the University of Cambridge to study mathematics, physics, and astronomy. By combining his interests in physics, mathematics, and astronomy, Newton was able to calculate how objects move and worked out a proof that showed the effect of gravity on the planets. Through his work, Newton determined the mathematical equations for the laws of motion.

One law that Newton discovered is called the law of universal gravitation. (We will discuss gravity later in this book.) Newton was able to confirm Galileo's experiments and showed mathematically why two falling objects will reach the ground at the same time even if one is heavier than the other.

One of the great discoveries of Newton's time is that mathematics can be used to describe events that happen in nature. For example, Newton was able to show that the force acting on an object is proportional to the mass of each object and inversely proportional to the square of the distance between them. The equation is:

$$F = G\frac{m_1\, m_2}{r^2}$$

where F is *gravitational force,* G is the *gravitational constant,* m_1 is the *mass of object 1,* m_2 is the *mass of object 2,* and r is the *distance between them.* (F, m_1, m_2, and r are called variables because the amount they stand for can change, or vary. G is called a constant because its amount stays the same in different equations. Inversely proportional means that as one variable increases in value, another decreases.)

How can this equation be used to explain Galileo's experiment? If we fill in the values for each variable and use m_1 for the mass of Earth, we can see that because the mass of Earth (m_1) is huge and the mass of each ball (m_2) is very tiny in comparison, when the mass of Earth is multiplied by the mass of the ball, the value of F won't be affected by the mass of the ball. Therefore, the mass of each ball can be ignored because it doesn't make any difference to the answer to the equation. In other words, the equation shows that the gravitational force on any object is the same regardless of its mass as long as its mass is much smaller than the mass of Earth. This means that any two objects will fall at the same rate even if one object is heavier than the other. The equation that expresses this is:

$$F = Gm \quad \text{(m=mass of Earth)} \quad \text{(for both balls)}$$

By using mathematics, Newton was able to prove Galileo's experiment.

1.4 Summary

- Physics is the study of how things move and behave in nature.

- The laws of physics are precise statements about how things behave.

- The laws of physics were determined using scientific investigation.

- Mathematics can be used to describe events that happen in nature.

1.5 Some Things to Think About

- What have you observed about objects that are or are not in motion?

- What do you think your life would be like if there were no physical laws?

- If you dropped a bowling ball and a golf ball off a high tower at exactly the same time, do you think they would hit the ground at the same time or different times? Why?

Chapter 2 Technology in Physics

2.1 Introduction

In a nutshell, physics explores the basic laws of nature. However simple this statement appears to be on the surface, digging deeper we find that physics is a complex aggregate of many different and specialized topics in many areas of science. As a result, physicists require many different types of tools and technology.

The tools physicists need for exploring atoms and subatomic particles are different from those used to study black holes in the farthest reaches of space. Physicists use electronics and computers to study force, speed, energy, gravitation, chaos theory, dark matter, and even game theory. Physicists also use mathematical models, fundamental constants, statistics, standards, and timekeeping to investigate the basic laws of nature. To explore all these areas, physicists have invented a wide variety of different tools and technologies that allow them to study the natural laws that govern how the world works.

In this chapter we will take a small snapshot of just some of the tools and advanced technology used by physicists. We will examine a few core technologies that are used to explore both small and large scale phenomena, and we will also explore a few specialized instruments used for measuring fundamental laws of physics such as force, speed, and energy.

2.2 Some Basic Physics Tools
Measuring Force

Imagine you are pushing a heavy wheelbarrow up a hill. How much force are you using? Do you think it takes more force to squeeze a rubber ball or a marshmallow? How can you tell? Is there a way to measure force?

When you push, pull, squeeze, or elevate an object, you are using force. Force is an action that changes the speed, shape, or position of an object. (We will learn more about force in Chapter 3.) A force gauge is an instrument that can be used to measure compression, tension, torque, and gravitational force which causes weight. There are several different types of force gauges. Two of these are mechanical force gauges and digital force gauges which are generally handheld instruments that can measure tensile force (pulling) and compression force (pushing). When testing objects, mechanical force gauges and digital force gauges can be held in the hand or mounted on a stand, depending on the requirements of the test being performed. The object or objects to be tested are attached to the force gauge which measures the pushing or pulling force being applied to the object.

A mechanical force gauge uses moving parts to measure the amount of force. A needle on a dial indicates how much force is being applied. A digital force gauge uses electronics to measure the amount of force being used and displays the results on a digital screen. Digital force gauges can quickly determine whether the holding strength of a connector or other object is strong enough that it won't break when being used. For example, if you were going to use a rope to lift a heavy box, before you hooked the box to the rope you might want to check how much force the rope can withstand and how much force the box will create on the rope!

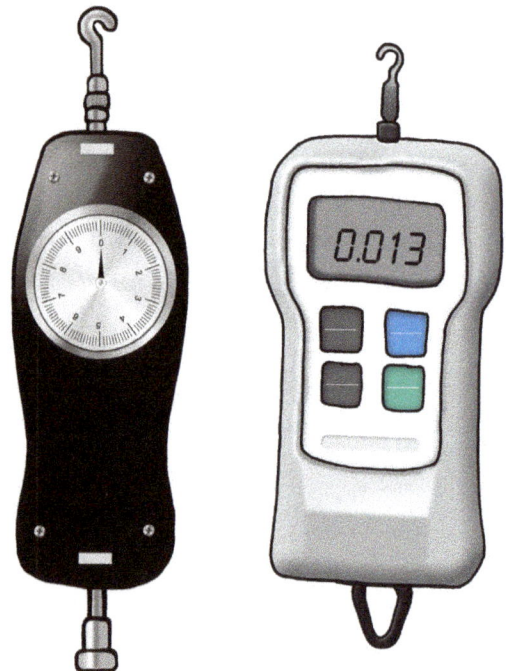

Torque gauges measure torque. Torque is the force that is exerted on an object when the object is being rotated around an axis. Torque gauges measure how much torque is being created. For example, when you use a wrench to attach pedals to your bike, you are applying torque with the wrench. A torque wrench, a wrench that contains a torque gauge, is a great way to determine the point at which you are applying just enough torque but not too much!

Measuring Speed

How fast can you run? How fast can a car go? How fast does a bullet go? How fast does an electron travel? How can you measure the speed of a runner, a car, a bullet, or an electron? As we will see in Chapter 7, speed is the distance traveled divided by time, so both distance and time need to be measured to determine speed.

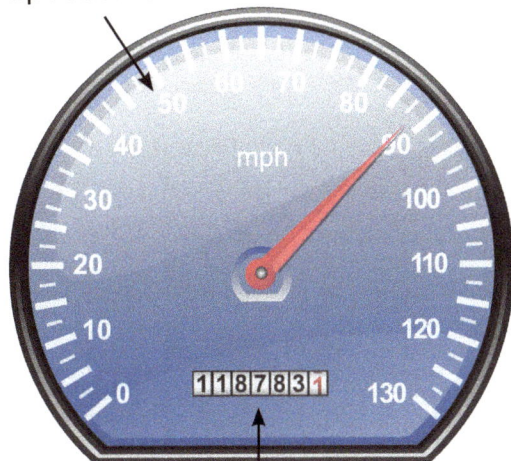

Speedometer

Odometer

In automobiles a speedometer is used to measure speed. A speedometer is a type of gauge that works from the car's driveshaft—the shaft used to turn the wheels. As the driveshaft turns, the speedometer measures how fast the driveshaft is rotating the wheels and uses this information to calculate the number of miles per hour being traveled. The odometer is the gauge that displays the distance traveled, which can be calculated by multiplying the number of wheel rotations by the size of the tire.

If you know the distance traveled, speed can be calculated by using a stopwatch. A stopwatch measures time and can be started and stopped at different intervals. Stopwatches can have a digital screen that displays a number or a dial with a needle that moves. Stopwatches can be used for measuring speed during a physics experiment and on the track field to tell how fast sprinters and distance runners are going.

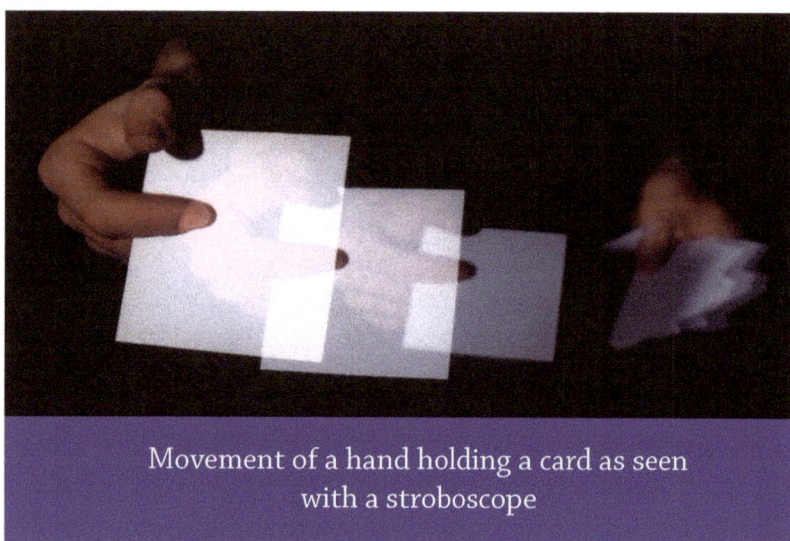

Movement of a hand holding a card as seen with a stroboscope

A strobe light or stroboscope is used to measure the speed of a rotating motor, fan, or other object. A strobe light flashes on and off very rapidly causing the movement of the object to appear as a series of snapshots rather than a continuous flow. This can produce the illusion of the object moving in slow motion. If a particular place on a motor or fan is marked with a dot, a strobe light or stroboscope can be used to help measure how fast the object is turning. The speed of rotation can then be adjusted to the desired speed.

Lasers are tools that emit a narrow beam of concentrated light and can be used to measure speed. If you've ever been in a car that was going over the speed limit and passed a parked police car only to find the police car soon following you with warning lights blazing, you just had your speed measured with a laser speed gun. A laser speed gun works by bouncing light off a moving object. Light is emitted from the speed gun in a narrow, concentrated beam. The light reaches the moving object, bounces back, and is detected by the speed gun.

The speed gun calculates the speed of the moving object by using the time required for the light to bounce back to it.

Measuring Energy

Tools for measuring energy are very important in physics. Energy can come in a variety of forms including heat energy, electrical energy, chemical energy, nuclear energy, mechanical energy, and light energy. We will learn more about energy in later chapters.

Thermometers are used to measure temperature. For many materials, heat energy is closely related to temperature. A bulb thermometer uses the expansion of materials such as mercury or alcohol to measure heat energy by displaying the amount of heat energy as temperature. Both alcohol and mercury will expand (become bigger) when heated and contract (become smaller) when cooled. By placing alcohol or mercury in a thin tube that has graduated markings, the temperature (which is related to heat energy) can be measured.

Electronic thermometers can also be used to measure temperature. An electronic thermometer contains a battery that creates an electric current, and it also contains a thermal sensor (heat sensor) called a thermistor. A thermistor is a component made of a special ceramic or polymer material that senses heat. This material is also a resistor that allows more or less electricity to flow through the thermistor according to the temperature being measured. A simple computer in the thermometer translates the amount of electrical flow into a temperature and displays this number on a screen. Electronic thermometers are used to measure the heat of an oven, the temperature of a car engine, and even the temperature of living things like animals and humans.

Voltmeters are used to measure electrical energy. A voltmeter measures the voltage difference between two points in an electric circuit. Voltage is a measure of electromotive force (EMF). A voltmeter has a needle gauge or digital display that shows the voltage difference between its two leads. Portable voltmeters have leads that can be attached to positive and negative terminals. For example, you can use a voltmeter to measure the voltage (potential electrical energy) inside a battery by placing one lead on the positive battery terminal and one on the negative terminal.

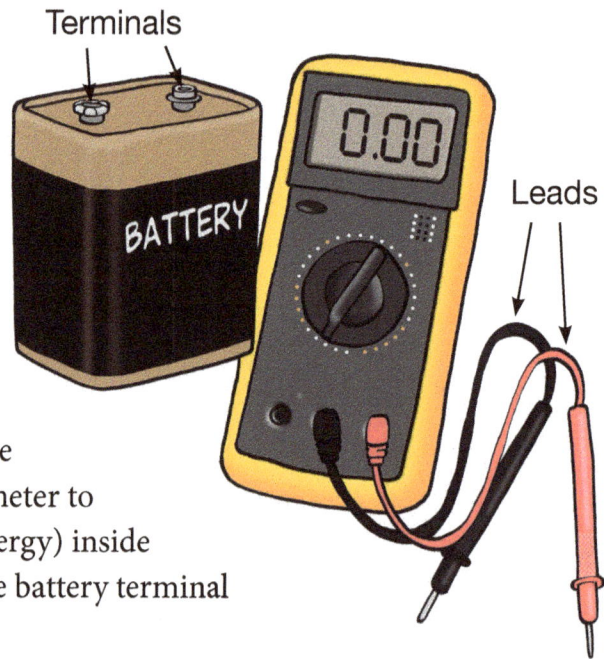

Voltmeters can often be switched between measuring voltage and measuring current. Current is the rate of flow of electrical charge between two points. Current and voltage are related but are not the same thing. We will learn more about voltage and current in Chapter 11.

2.3 Mathematics

One of the most amazing and beautiful features of physics is that physical laws can be expressed precisely using mathematics. The use of mathematics is an essential tool in physics. Without mathematical calculations we would not be able to see distant stars, explore the depths of the oceans, land a probe on a comet, or visualize atoms and molecules.

One of the most basic mathematical tools used in physics is geometry. Geometry uses distances and angles to describe the exact relationship of objects to each other. The mathematics behind geometry is largely attributed to the Greek mathematician Euclid (circa 325 BCE-circa 265 BCE). Euclid's geometry, called Euclidean geometry, is based on a set of axioms. An axiom is a statement that is used as a starting point, premise, or postulate. For example, one of Euclid's axioms states that "any line segment can be extended indefinitely in a straight line." As we will see in Chapter 8, *Non-linear (Curved) Motion,* we can use this axiom to calculate the velocity of a stone that has been thrown into the air.

Algebra is also an important mathematical tool for physics. Algebra is a type of mathematics that uses symbols in arithmetic calculations. Some historians think that algebra was likely developed by the Persian mathematician al-Khwarizmi (circa 780-circa 850 CE) and then further developed by the French mathematician Francois Viete (1540-1603 CE) in the late 16th century.

EUCLID
Circa 325-265 BCE

Algebra is used to relate and solve a variety of relationships in physics. For example, the speed of an object is defined as:

$$speed = \frac{distance}{time}$$

If you want to calculate the speed of a car going from San Francisco to Los Angeles, you can plug in the values for "distance" and "time."

$$speed = \frac{382 \text{ miles}}{6 \text{ hours}}$$
$$speed = 64 \text{ miles per hour}$$

But what if you know you can only go 50 miles per hour? How can you find out how long it will take?

With algebra you can generalize the equation using mathematical symbols. Then you can rearrange the equation to solve for the number of hours.

Using mathematical symbols the equation now becomes:

$$s = \frac{d}{t}$$

where "s" represents "speed", "d" represents "distance" and "t" represents "time." This is a simple algebraic equation.

Applying the rules of algebra, we can rearrange the symbols and solve for time "t":

$$t = \frac{d}{s}$$

Plugging in the values for distance (d) and speed (s) you have:

$$t = \frac{382 \text{ miles}}{50 \text{ miles per hour}}$$
$$t = 7.64 \text{ hours}$$

More complex physics requires more complex mathematical tools. Trigonometry, differential equations, complex algebra, and vector calculus are all complex mathematical tools used in physics. Without the use of mathematics, physics would still be considered a "philosophy" rather than a science in which exact quantities can be measured, evaluated, and calculated.

2.4 Electronics

Discoveries about how electrical currents work have made many advancements in science possible. The development of electronic equipment for physics is a great example of science influencing technology and technology influencing science.

Electricity is the set of phenomena that arises from electrical charges and movement of electrical charge. Electric charge in the form of electrons can move through metals, semi-conductor materials, and even a vacuum and can produce electrical power, electric potential, and electromagnetic fields.

The nature of moving and static electric charge was an area of active investigation for many early scholars. People in ancient times knew that certain objects could produce an electric shock, but they weren't able to determine why this happened. It wasn't until the late 17th century that theories about atoms were developed and explored. Although experiments were performed to investigate electrical currents, the connection between electrons and atoms wasn't discovered until much later.

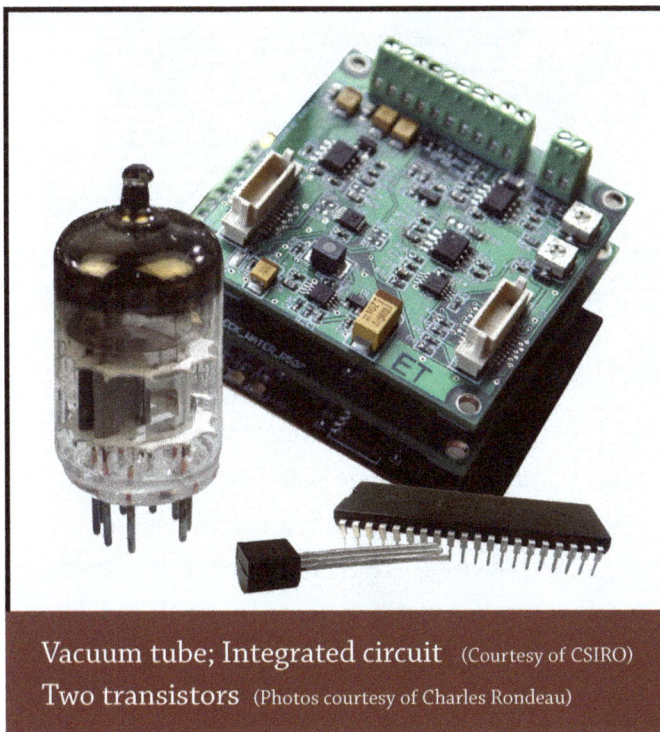

Vacuum tube; Integrated circuit (Courtesy of CSIRO)
Two transistors (Photos courtesy of Charles Rondeau)

As information about atoms, electrons, protons, and neutrons became available, understanding of electrical phenomena increased. Experiments using electrical currents led to the development of complex technological components including the vacuum tube, transistor, and integrated circuit. As this technology continued to be developed, more physics theories could be explored. The cycle between technology advancing science and science advancing technology continues today.

A vacuum tube is a glass tube or bulb that controls electric current through a vacuum that is sealed inside the tube. The invention of the vacuum tube made it possible to manipulate, amplify, and transmit electrical energy. The vacuum tube was used in early radio communications and made development of radio possible. Vacuum tubes were also used in early computers but were replaced when the transistor was invented. Transistors function like the vacuum tube, manipulating, amplifying, and transmitting electrical energy, but transistors are small, weigh less, and require less power to run. Integrated circuits regulate electrical energy by combining transistors and other electrical components such as capacitors that store electric charge and resistors that limit the flow of electric current. Equipment is called electronic when it contains these kinds of small electrical components.

Electronic instruments use the movement of electric charge through an electrical circuit to create a variety of different outcomes. Often electronics are used to convert energy from one form to another. For example, a flashlight uses electronic components to convert chemical energy into light energy. Electronics and electric circuits can also be used to convert chemical energy to movement, sound, or heat energy. Electronics and electric circuits are used in many toys, tools, appliances, computers, and scientific instruments.

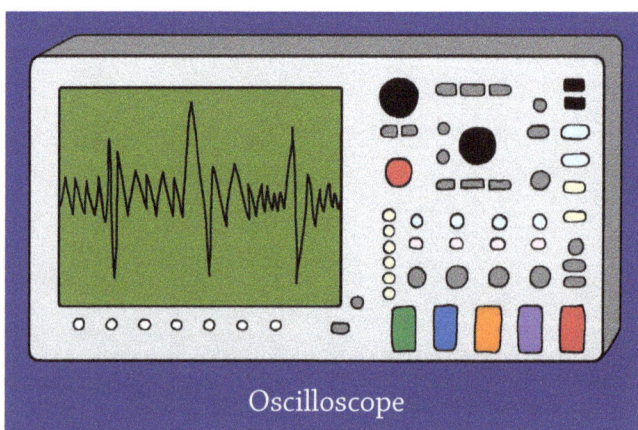

Oscilloscope

Electronics found in many modern physics labs include simple instruments, such as stopwatches that measure time, to more advanced instruments, such as those that detect the number of atoms on a silicon surface. Physicists use electronic oscilloscopes to observe how voltage signals vary with time, voltmeters to measure the amount of voltage, and spectrometers to measure light. The development of electronics has made all of these instruments possible.

2.5 Computers

Computers play a vital role in physics labs today. Most electronic equipment can be attached to a computer. Computers allow physicists to quickly collect, modify, and analyze data. Computers also allow physicists to create computer simulations and models of physical data. These simulations and models help scientists understand and predict a variety of physical phenomena. For example, physicists and engineers can use computer models to explore airplane designs and can use a flight simulator to find out how an airplane might react during extreme weather.

A robot can work under dangerous conditions

Courtesy of DARPA

Computers and computer models are also used in the field of robotics. Exploring how to convert electrical energy into mechanical energy is important in robotics. Having a robot process information as easily as a human is one of the many goals of robotics. With computers and computer models physicists can determine how to create robotic movements that copy human legs for walking and the human hand for grasping objects.

Computers are used in optics, laser, and microscopy labs. The fine adjustments needed to control a scanning tunneling microscope (STM) are generated by a computer that sends electrical signals to a piezoelectric crystal that moves a tiny sample by very small amounts. Because such small movements are possible, individual atoms can be seen.

2.6 CERN

By putting new theories and discoveries in physics together and using the most advanced technology and mathematics, scientists can now explore the very fundamental structure of the universe. At the laboratory that straddles the border of France and Switzerland, the European Council for Nuclear Research (CERN) has assembled a technological masterpiece for probing the basic constituents of matter and the forces that act on them.

Aerial view of CERN - circle shows the location of the underground Large Hadron Collider

Courtesy of CERN/ Maximilien Brice, photographer

At CERN, whose name comes from the acronym for the French "Conseil Européen pour la Recherche Nucléaire," physicists are exploring theories about the universe, such as antiprotons (protons with a negative charge), dark matter (invisible, as yet unknown matter that makes up most of the universe), particles called quarks and leptons that are thought to be smaller than protons and neutrons, and a very special particle called the Higgs boson that physicists think may be what gives matter its mass.

Thousands of scientists and engineers from many countries all over the world have worked with CERN to build and operate a huge instrument called the Large Hadron Collider (LHC) that explores what happens when parts of atoms (particles) are smashed together at very high speeds. The Large Hadron Collider is a particle accelerator in which beams of particles shoot through the collider's tunnel — a structure that is underground, donut shaped, and 27 kilometers (17 miles) in circumference.

The Large Hadron Collider contains two very sophisticated pieces of equipment—an accelerator and a detector. The accelerator creates two beams of protons from bottled hydrogen gas and pushes the protons to go very fast—so fast that they are traveling at nearly the speed of light, making 11,000 circuits per second around the tunnel! The two beams are traveling in opposite directions, and once the protons are moving at nearly the speed of light, the two beams of protons are made to collide with each other. When particles smash into each other, a detector gathers up all the information. There are four detectors that determine if the protons have smashed into other particles, and if so, how fast they were going, how big they were, and their charge. Data from the detectors is sent to the CERN Data Centre and from there it is sent through the Worldwide LHC Computing Grid to labs in many different countries, giving thousands of physicists access to the data.

Inside the Large Hadron Collider tunnel

Courtesy of CERN/ Maximilien Brice, photographer

Physicists can use the Large Hadron Collider to study a specialized field of physics called particle physics. Using data from the LHC, physicists can study how cosmic rays affect Earth's atmosphere and cloud formation, how antiprotons interact with living things, and how particles are generated by the Sun.

2.7 Summary

- Physics includes many different and specialized topics and requires many different types of tools and technology.

- Some basic physics tools allow physicists to study force, speed, and energy.

- The use of mathematics is an essential tool in physics, and physical laws can be described exactly through mathematics.

- The study of physics uses advanced, specialized electronic equipment along with simpler instruments.

- Electronics and computers play a vital role in physics.

- CERN's Large Hadron Collider is helping physicists explore the fundamental structure of the universe.

2.8 Some Things to Think About

- What area of physics would you most like to explore? Why?
  - Force • Speed • Energy • Gravity • Chaos theory • Game theory • Black holes • Planetary orbits • Another area (describe)

- What tools of physics have you used or seen used? What did you use them for?

- Why do you think mathematics is important for physics?

- Do you think the discovery of geometry led to advances in physics, geology, and astronomy? Why or why not?

- Look around your house. How many devices can you find that are electronic? Do you think there are any that use electricity or batteries but are not electronic? Why or why not?

- How do you use computers and computerized devices in your life today?

- What do you think your life would be like if all the computers suddenly disappeared?

- How do you think the invention of the computer affected discoveries in physics? Why?

- Do you think physicists will eventually be able to discover all the particles that exist in the universe? Why or why not?
  Do you think they will be able to tell when they have found the smallest particle that exists? Why or why not?

Chapter 3 Force, Energy, and Work

3.1 Introduction

What is energy? When your mom says, "I am out of energy," what does she mean?

Energy is actually defined as the *ability to do work*. The term *work*, as used in physics, describes what happens when a force moves an object. Your mom may need to rest to get back the energy she needs in order to have the force to move herself and other objects *(do work)*.

This can seem a little confusing, so let's look at force, work, and energy in more detail.

3.2 Force

What is force? Have you ever dropped an egg on the floor? What happened? Probably you heard a noise and noticed that the egg was no longer available for your cake. In fact, you probably had to clean up a sticky mess. What happened to the egg? Why did it break? It broke because of force. The egg hit the floor with enough *force* to break it open.

Have you ever pushed on a heavy door that just wouldn't open? Did the door feel like it was pushing you back? When we push on a door, we apply a force to the door to open it or to move it. The door pushes back. The same thing happens when we pull on the door; the door pulls back. Both the pushing on the door and the pulling on the door are forces. A *force* is...

> something that changes the *position, shape, or speed* of an object.

There are many different sources of force. You experience one source of force every day, all day long. That is the force of gravity. The Earth is the source of the gravitational force you experience. It pulls on you and makes you, and everything else, stick to the ground.

The force of gravity is actually exerted by every object. You also are a source of gravitational force, and you pull on the Earth at the same time the Earth pulls on you. However, because you are so much smaller than the Earth, your gravitational force is very small compared to the gravitational force of the Earth. So, instead of dragging the Earth with you out into space, the Earth keeps you tightly stuck on its surface. In fact, all of the planets exert gravitational force. They pull and push on each other, and as a result, their distances from the Sun and their orbits around it are balanced and stay the same.

3.3 Balanced Forces

An object that is not moving has balanced forces. For example, a toy sitting motionless on your bookshelf is actually applying a force downward toward the shelf, and the shelf is applying a force upward toward the toy. The forces are *balanced*; they cancel each other out, so the toy does not move.

Another way to look at this is to consider what happens if you and your friend are pulling a rope in opposite directions. If you both pull with equal strength and

Toy pushes down on shelf

Shelf pushes up on toy

neither of you can move the other, then the forces with which you pull are equal. The forces are *balanced*. You both remain motionless.

Balanced forces can also occur with objects that are moving. For example, an air hockey puck slides gracefully, at the same speed, across a hockey table until it is struck with an opponent's paddle. As it is moving and as it is at constant speed, the forces between the puck and the table are balanced. This happens with anything that slides, such as snow skis, ice skates, or even magnetic trains!

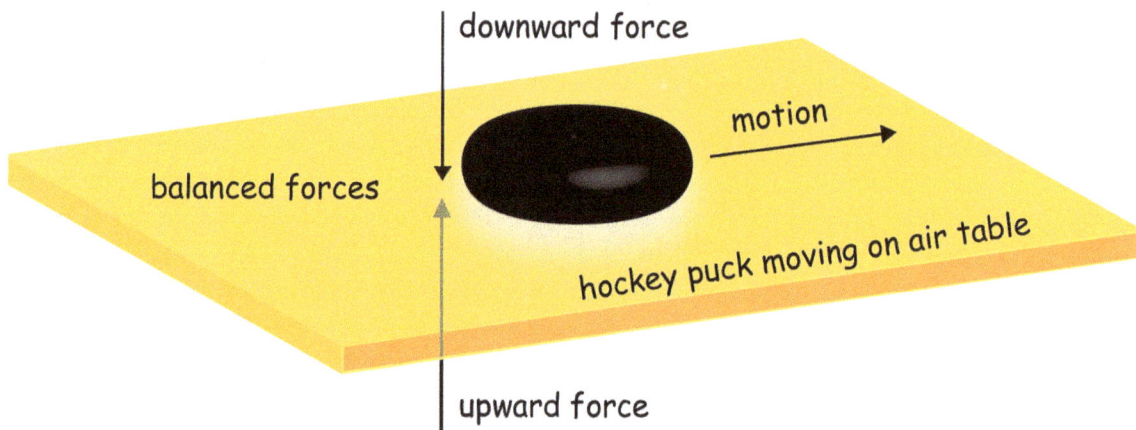

downward force

motion

balanced forces

upward force

hockey puck moving on air table

3.4 Unbalanced Forces

If the forces are unbalanced, that is, one force is greater than the other, the object will move. As long as the force keeps acting on the object, the object keeps moving faster. If the object keeps going faster and faster, it is said to accelerate. Unbalanced forces always cause acceleration.

Suppose that when you and your friend are both pulling on the rope, you suddenly pull less hard on your end. What happens? Your friend keeps his force the same, but because you are pulling less hard, BAM! He's in the puddle! Why?

Your force decreased, and your friend's force became greater than your force. These unbalanced forces caused him to fall backward into the puddle. When the forces were equal, you and your friend did not move. As your force decreased (you pulled less hard), your friend began to move. In other words, he went from no speed (standing still), to some greater speed (falling in the puddle).

This change in speed is called acceleration, and this acceleration was caused by a force. In this case, your pulling less strongly on your end of the rope caused your friend to accelerate into the puddle.

3.5 Work

What is work? You probably hear comments like, "I am late for work," by your dad, or "I have too much work," exclaimed by your mom. You might think that work is a very grown-up thing that causes lots of stress, and your parents might agree. But in physics, work is something very simple. *Work* is the result of a force moving an object a certain distance.

When force is used to move an object a given distance, work has been done on that object. The amount of work done is calculated by

multiplying the force times the distance the object has traveled. This can be expressed as a mathematical equation:

$$w = d \times f$$

where w stands for *work*, d stands for *distance*, and f stands for *force*.

For example, as the expression on the face of a weight lifter shows, a tremendous amount of work is needed to lift a heavy barbell from its resting position on the ground to its final position above the weight lifter's head. The amount of work the weight lifter does is proportional to the distance he has to lift the barbell. *Proportional* means that work and distance are related; if there is twice as much distance, the weight lifter does twice as much work.

For example, a very short weight lifter would have to do less work to get the bar above his head than a very tall weight lifter. If the short weight lifter were half the height of the tall weight lifter, then he would do exactly half the amount of work.

3.6 Energy

When work has been done, and forces have been used to do that work, energy has been used. It's hard to define energy exactly, but one thing energy *does* is to give objects the ability to do work. Take a look at the weight lifter we studied in the last section. When the barbell is on the ground, it requires the force of pulling up on the barbell to lift it above the weight lifter's head. When this happens, work has been done. But where did the weight lifter get what he needs to lift the barbell? Wheaties! Yes! The weight lifter had to have energy in his body to be able to use his muscles to do the work of lifting the barbell above his head. By eating food, living things get a type of energy needed to do work.

There are actually different kinds of energy because there are different ways to do work. The different types of energy are given different names. A few of these different types of energy are potential energy, kinetic energy, and heat energy. We will look at some different types of energy in more detail in later chapters.

3.7 Summary

● A force is something that changes the position, shape, or speed of an object.

● Forces can be balanced or unbalanced. Objects that are not moving, or objects that are moving at constant speed, have balanced forces. Objects that are accelerating have unbalanced forces.

● Energy is hard to define, but it gives objects the ability to do work.

● Work = distance x force. This means that, for example, twice the distance gives twice the work for the same force.

3.8 Some Things to Think About

● What do energy, force, and work mean to you?

● What do you think would happen if Earth no longer had gravity?

● Make a list of some examples of balanced forces.

● Make another list that has some examples of unbalanced forces.

● What examples of acceleration can you think of?

● Do you think you'd be doing the same amount of work if you carried a box of books up one flight of stairs and then carried a box of books weighing the same amount up two flights of stairs? Why or why not?

● How has physics changed your idea of what energy is?

Chapter 4 Potential and Kinetic Energy

4.1 Potential Energy

What is potential energy? You've probably heard the word potential used before. For example, you may have heard someone say, "He's got potential," or "The tropical storm has the potential to become a hurricane." In both of these statements, the word *potential* refers to something that has the capacity to happen or become. "He's got potential" simply means that someone has the possibility of becoming something like a great basketball player or leader in the future but isn't one right now. The tropical storm may become a hurricane, but it isn't right now. It only has the potential to become one. Recall that energy is used to do work. Simply put, *potential energy* is energy that has the potential to do work.

Potential energy is a type of energy often called stored energy. An example of an object with potential energy is a book on a table. It may not seem like the book can do work, but because the book is not on the floor but is raised, it has the potential to fall off the table.

When the book falls off the table, it strikes the floor with a *force*. This force could be used to crack open a peanut, smash a marshmallow, or make a big noise. The book can use the potential energy to do work.

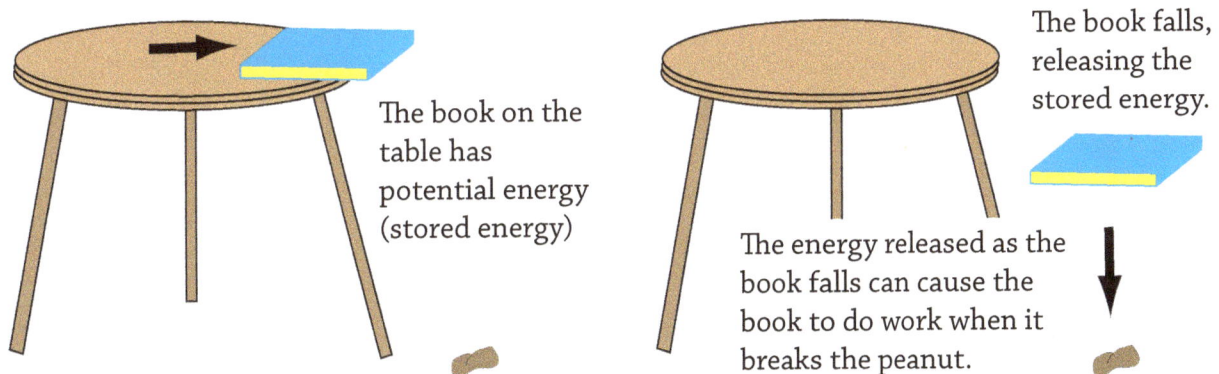

The book on the table has potential energy (stored energy)

The book falls, releasing the stored energy.

The energy released as the book falls can cause the book to do work when it breaks the peanut.

This type of potential energy is called gravitational potential energy because the force of gravity is required to bring the book from its elevated position (on the table) to its final position (on the floor). The amount of gravitational potential energy of an object equals the amount of work that was needed to lift the object in the first place.

The amount of gravitational potential energy can be calculated by multiplying the weight of the object by the height:

gravitational potential energy (GPE) = weight x height

For example, if the book on the table is 1 meter (3.28 feet) above the floor and it weighs 1 kilogram (2.2 pounds), the gravitational potential energy (GPE) is:

GPE = 1 meter x 1 kilogram (3.28 feet x 2.2 pounds)

or

GPE = 1 kilogram-meter (7.2 foot-pounds)

4.2 A Note About Units

What is a unit? In physics, a *unit* is simply the name given to a type of measurement. For example, to measure your height you might stand next to a wall and have your mom mark a place on the wall by the top of your head. Using a ruler, she can then measure how tall you are by putting one end of the ruler on the floor and the other end on the mark on the wall. Your height might be something like 1.25 meters (4 feet, 2 inches). *Meters* and *feet* and *inches* are called *units*.

Meters, feet, and inches measure how long something is, but other units, like kilograms and pounds, may tell us how much something weighs, or like liters and gallons, how much liquid something can hold. Time also has units, like hours, minutes, and seconds. It tells us how long something takes to happen; for instance, how long it might take for an egg to reach the ground if it is dropped from a tall building.

In the United States, we often use what are called British units, for example, feet and inches. But most scientists use metric units. Metric units are usually easier to work with than British units because they can be evenly divided by 10. British units are usually converted to metric units when used in science. The table shows some units in both metric and British.

British Units		Metric Units	
	1 inch	10 millimeters	1 centimeter
12 inches	1 foot	100 centimeters	1 meter
5280 feet	1 mile	1000 meters	1 kilometer

4.3 Types of Potential Energy

There are actually several different types of potential energy. We already saw gravitational potential energy, which is energy associated with the position of an object. Other types of potential energy include nuclear potential energy, elastic or strain potential energy, chemical potential energy, and several others.

Nuclear potential energy is the energy that is stored in an atom. Nuclear reactors use the nuclear potential energy that is stored in uranium atoms to heat water, which can then be used to make electrical energy. Nuclear reactors can provide electricity for very large communities and even whole countries!

Elastic or strain potential energy is the energy stored in an extended rubber band or a compressed spring. Chemical potential energy is the energy that is stored in molecules, such as that found in batteries, fuels, or foods.

4.4 Energy Is Converted

What happens to the potential energy of the book, the battery, or the rubber band once the energy is released? Is it still potential energy? No. The potential energy of the book, the battery, and the rubber band have all been released and converted into another type of energy. It is important to know that:

Potential energy is useful (can do work) only when it has been converted into another form of energy.

Can you think of other uses for batteries? Tree decorations perhaps? Or maybe a nice hood ornament? Not really. In fact, batteries are useless unless their potential energy is converted — for instance, to light a flashlight or power a CD player. When a battery is used to power a CD player or a flashlight, the chemical potential energy inside the battery is released by chemical reactions and converted to electrical energy. The electrical energy can then be converted into light energy in the flashlight or mechanical energy in the CD player.

4.5 Kinetic Energy

We saw in the last section that potential energy must be converted into another form of energy before it can do work. What kind of energy is it converted into? When the book was dropped from the table, the gravitational potential energy had to first be converted into kinetic energy before it could do work on the peanut.

What is kinetic energy? The word kinetic comes from the Greek word *kinetikos*, which means "putting into motion," and kinetic energy is the energy associated with things that are moving.

The potential energy of the book on the table is converted into kinetic energy when the book falls — that is, while it is moving toward the floor. The book has no kinetic energy as it sits on the table, only potential energy. When the book is moved from the table and begins to fall, the potential energy is converted into kinetic energy. The farther it falls, the more kinetic energy it gains and the more potential energy it loses.

By the time it hits the floor, all of the potential energy has been converted into kinetic energy. The total amount of energy has not changed — only the form of energy. Physicists say that the total energy is conserved. That is, all of the potential energy has been converted into another form of energy. Energy is never lost — only converted. We will learn more about the conservation of energy in the following chapter.

How much kinetic energy does the book have? It depends. The kinetic energy of an object depends on two things — one is the *mass* of the object, and the other is the *speed* of the object.

What we need to remember about kinetic energy is the following:

For a certain speed, the more mass an object has, the more kinetic energy it has;

and

For a certain mass, the more speed an object has, the more kinetic energy it has.

Therefore, a heavy book will have more kinetic energy than a lighter book moving at the same speed. Also, a book that is thrown will have more kinetic energy than a book that is dropped.

4.6 Kinetic Energy and Work

We already saw in the last chapter that energy is the ability to do work. When a rubber band is stretched across the prongs of a slingshot, it has *elastic potential energy*. When the rubber band is released, the elastic potential energy is transferred to the pellet in the slingshot as the pellet is propelled toward the target. The pellet now has *kinetic energy*. All, or almost all, of the potential energy that was in the slingshot is now kinetic energy in the pellet.

What happens to the kinetic energy in the pellet when it hits the target? The kinetic energy is converted to other forms of energy, such as heat and sound. As a result, the energy is transferred to the target in the form of work as it pushes on the target.

We say that:

> *The pellet is doing work on the target.*

Out of my way! I've got work to do!

4.7 Summary

● Potential energy is energy that has the potential to do work.

● A book on a table has gravitational potential energy.

● The energy in a stretched rubber band is called elastic potential energy.

● Kinetic energy is the energy of motion.

● Potential energy can do work only when it is converted into another form of energy, such as kinetic energy.

4.8 Some Things to Think About

● Make a list of objects that you think have gravitational potential energy and explain why they do.

● Take a ruler that has British units on one side and metric on the other. Measure the width of a sheet of paper and write down this measurement in British units and in metric units. Now divide each of these numbers by 3. Which calculation is easier?

● Look at the units of measure on a glass measuring cup. Do you see both British and metric units? If so, pour a glass of water into the measuring cup and compare the British unit and metric unit readings. Do you think one system would be easier to use for scientific measurements? Why?

● Why do you think potential energy is important?

● What are some examples of potential energy being converted into energy that is used?

● Name some ways that you use kinetic energy every day.

● What are some examples of potential energy being converted to kinetic energy that then does work?

Chapter 5 Conservation of Energy

5.1 Introduction

Recall the different types of energy we have looked at, such as *potential energy, kinetic energy, chemical energy,* and *electrical energy.* We have seen how one form of energy can be converted into another form of energy. The *gravitational potential energy* of a toy car can be converted into *kinetic energy* as it rolls down a ramp to smash a banana.

We have seen how *chemical potential energy* inside a battery can be converted to *light energy* in a flashlight and how *mechanical energy* can be converted to sound in a CD player. If we recall from biology how plants make food, we can see that chloroplasts convert *light energy* from the Sun into *chemical energy* in leaves. From chemistry, we learn that when we eat food, such as carbohydrates, we get *chemical energy* for our bodies. When we lift a weight or run a race, this *chemical energy* is converted into *mechanical energy.*

5.2 Energy Is Conserved

In all of these processes, energy is neither created nor destroyed. Energy is simply converted from one form to another. In fact, energy cannot be created or destroyed, but only converted. There is a fundamental law of physics, called the law of conservation of energy, which states that energy is conserved. This simply means that the total amount of energy we convert to other forms of energy does not increase or decrease — it stays the same. In fact, the whole universe has the same amount of energy today that it had ten or even twenty years ago. It will have the same amount of energy tomorrow and the next day that it has today! Even a hundred or a thousand years from now, the energy in the universe will stay the same. Energy is conserved.

How is energy conserved? In the experiment for Chapter 4, we saw how gravitational potential energy (GPE) was converted into kinetic energy (KE) when a toy car was used to smash a banana. The energy of this system, called total energy, is equal to GPE plus KE. We saw that as the toy car rolled down the ramp, it lost GPE and gained KE. But what happened to the total energy? Did it change too? No, in fact the total energy stayed the same. We can see how this happens if we look at both the GPE and KE at several places on the ramp.

Imagine that when the car is at the top of the ramp it has 100 joules (joules are a unit of energy) of GPE. Because it is not moving, it has no KE. When the car is halfway down the ramp, it has lost half of its GPE, but it has gained KE because it is moving. In fact, it has gained the same amount of KE that has been lost as GPE. Just before the car hits the bottom of the ramp, it has lost all of the GPE, but has gained more KE. The total energy (GPE + KE) remains the same at each point. The total energy does not change. This is what is meant by conservation of energy.

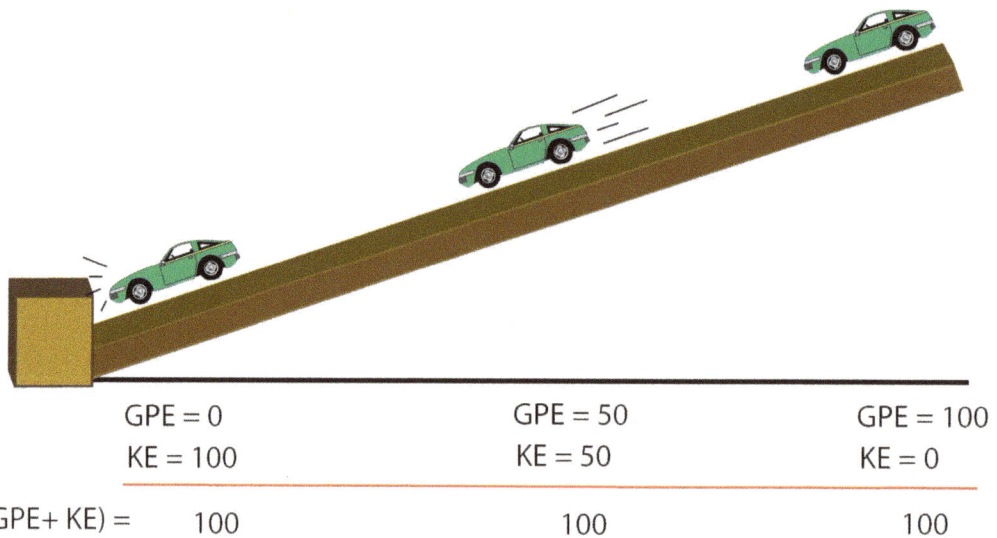

	GPE = 0	GPE = 50	GPE = 100
	KE = 100	KE = 50	KE = 0
Total Energy (GPE+ KE) =	100	100	100

5.3 Usable Energy

I BELIEVE ALL OF THE USABLE CHEMICAL ENERGY HAS BEEN CONVERTED INTO LIGHT AND HEAT.

Why is it that we often hear of an energy crisis or an energy shortage? Why are we concerned about fossil fuels or ways to save electricity? Why are we told not to waste energy? If we always have the same amount of energy, why do we care if we convert all of the chemical energy in a flashlight into light energy?

We care because not all energy is usable energy. That is, we cannot convert all of the energy into a form that we can use. For example, when a flashlight is left turned on behind the sofa, the stored chemical energy gets converted into light energy and heat energy. The usable energy in the flashlight is "lost" when all of the stored chemical energy has been converted into light and heat energy. We can't use it anymore. In fact, we have to throw the batteries away and get new ones! But the energy isn't gone, it has just been converted into a form that we can't use. So, when someone talks about an energy crisis, they mean that all of the *usable energy* is disappearing because it is being converted into unusable forms of energy, like heat energy.

5.4 Energy Sources

What are some of the forms of energy we use? We've already seen that batteries store chemical energy, but what about other forms of energy? Where do we get gasoline, electricity, and natural gas?

Some of the energy we use comes from fossil fuels, such as oil, natural gas, and coal. Fossil fuels are formed from plants and animals that died a very long time ago and have been subjected to extreme heat and pressure.

Plant fossil

When a plant or an animal dies, the tissues and cells that the plant or animal is made of decompose or break down into smaller pieces. One of the smallest pieces that all living things break down into is the carbon atom. Sometimes, if the conditions are just right, the carbon from the dead plants and animals combines with hydrogen and turns into oil, coal, or natural gas. These are called fossil fuels, or hydrocarbons. Hydrocarbons contain chemical potential energy. When hydrocarbons burn, the molecules combine with oxygen in a chemical reaction that converts chemical potential energy to other forms of energy.

Oil is found in underground rocks and is called crude oil or petroleum. Certain types of rocks, such as sandstone and limestone, are porous and hold oil in reservoirs, which are places where the oil collects and is stored. Porous rocks contain lots of empty spaces, or pores, and when the pores connect to each other, oil can flow through the rock. When these porous rocks are surrounded with nonporous rocks that the oil can't travel through, the oil will accumulate in reservoirs within the porous rock. Oil can be removed from a reservoir by drilling a hole deep into the rock and pumping the oil out.

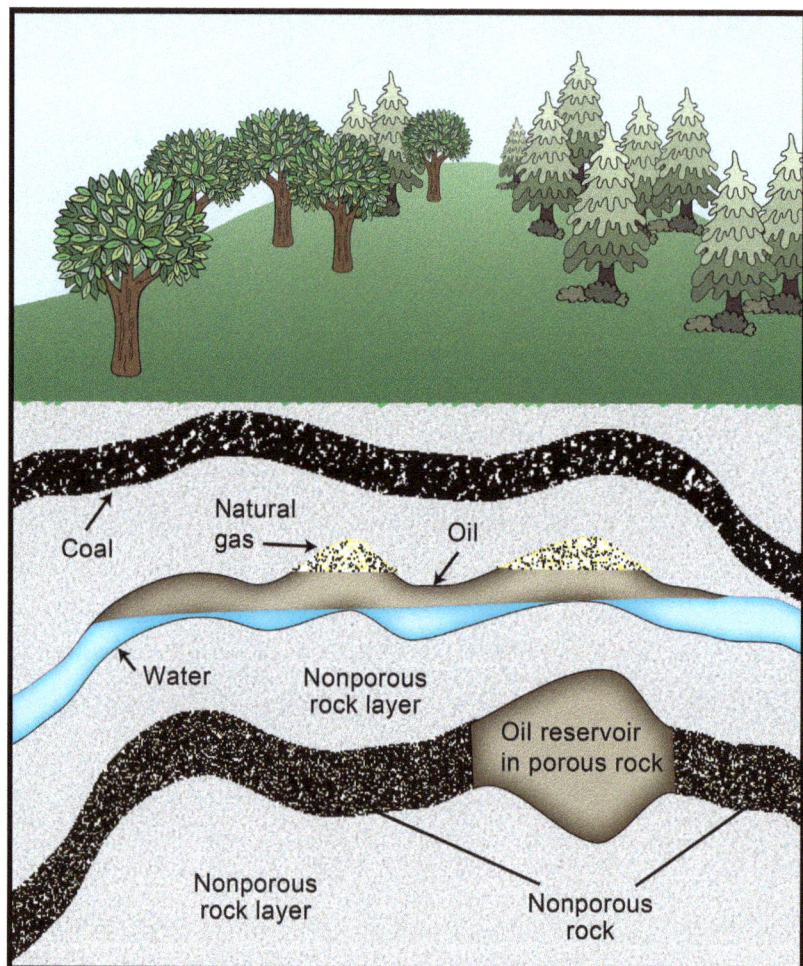

Crude oil is refined into different products, such as gasoline, kerosene, and jet fuel for cars, trucks, planes, and heating buildings. Petroleum is also used to produce substances such as plastics, waxes, fibers, and dyes. The chemical potential energy in petroleum fuels is converted into heat, light, mechanical, kinetic, and electrical energy.

Like oil, natural gas is formed from tiny plants and animals that have died, decomposed, and fossilized, and natural gas is often found in the same reservoirs as oil. Natural gas is made mostly of methane, which is a carbon atom with four hydrogens attached. Since it is lighter than oil, natural gas is found above the oil in a reservoir. Chemical reactions are used to convert the chemical potential energy in natural gas into heat, light, mechanical, kinetic, and electrical energy.

Coal is another fossil fuel found underground. It is formed from decaying plants that have fossilized and is found in layers, or seams. Coal is hard and cannot be pumped out of the ground like oil but must be removed by digging, or mining. Large holes or mine shafts, together with tunnels, are dug deep into the ground. Both miners and digging equipment can then get inside the ground to remove the coal.

Coal was the main fossil fuel up until the early 1900s. It was used to power steam engines and factories and to generate heat for making steel and iron. Today, coal is mostly used to generate electricity.

Coal miner underground

Although we have many uses for fossil fuels and have become dependent on them, in the process of burning, fossil fuels give off chemical by-products that can be harmful to the environment, and as fossil fuels are used, their potential energy is eventually converted to unusable energy. Fossil fuels are considered a nonrenewable energy source because they form so slowly that new fossil fuels won't be available to take the place of fossil fuels that have been mined. The supply of fossil fuels available today can be used up, and as the easily mined sources are removed from the Earth, fossil fuels become increasingly more difficult to mine. In addition, the mining, drilling, and processing of fossil fuels uses a great deal of energy.

There are energy sources other than fossil fuels that we can use for powering cars, for electricity in our homes, and for other uses. Two important sources of usable energy are

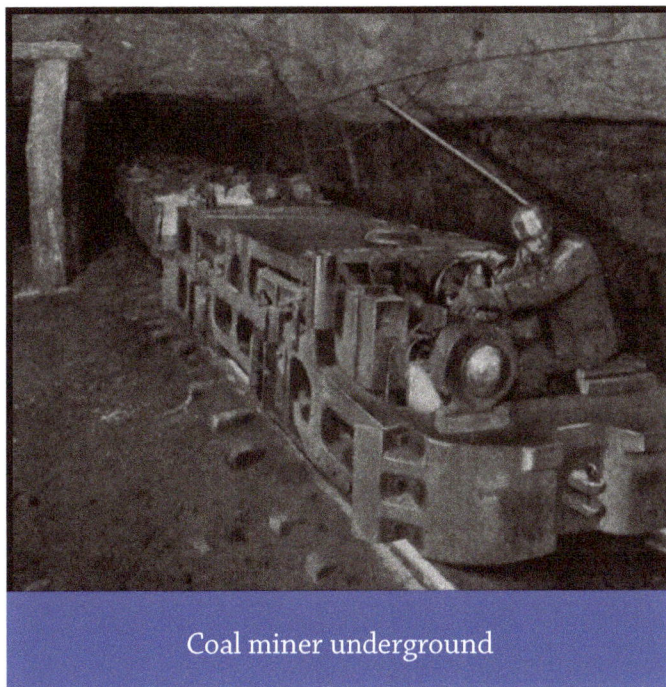

wind and water, which are considered to be renewable energy sources. This means that, unlike coal and oil, wind and water are continuously circulating rather than being used up. Heat from the Sun creates more wind, and water circulates around the Earth in the water cycle. Renewable energy sources don't create the pollution problems that the burning of fossil fuels does, and they are not going to run out.

Windmills harvest energy from the wind and have been used for centuries to pump water from the ground. Even today, it is common for windmills to bring groundwater to the surface for crops on farms and for cattle on ranches. Today's large windmills are called wind turbines and are being used to generate electrical energy. Earth's atmosphere is heated unevenly by the Sun, creating the kinetic energy of winds as air moves from high pressure to low pressure areas. As the wind hits the blades of a wind turbine, the gravitational potential energy of the blades is converted to kinetic energy as the blades move. The kinetic energy of the blades is converted to mechanical energy in an attached generator. The spinning of the generator converts kinetic and mechanical energy to electricity.

Flowing water in rivers is used in many places as a source of energy. A dam is built in a river to create a large lake, or reservoir, behind the dam. The water in the reservoir is higher than that of the river, giving it gravitational potential energy. As the water falls down from the reservoir through a passageway in the dam, the water's gravitational potential energy is converted to kinetic energy and has enough force to turn a turbine, creating mechanical energy which in turn is used to generate electrical energy. The reservoir is refilled by rain or flowing rivers, making water a renewable energy source.

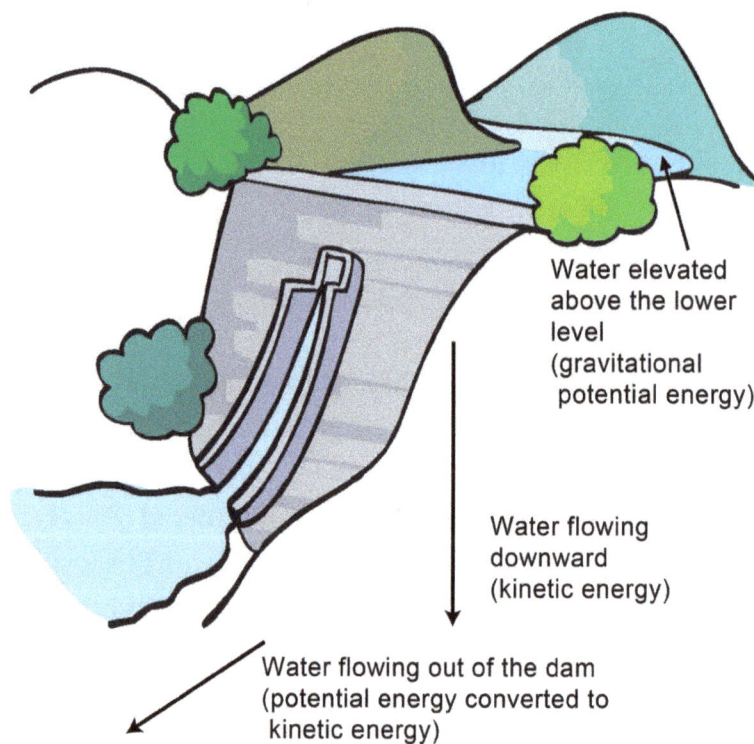

Water elevated above the lower level (gravitational potential energy)

Water flowing downward (kinetic energy)

Water flowing out of the dam (potential energy converted to kinetic energy)

The Sun is a very important source of energy for plants. Ultimately, all the energy contained in fossil fuels came from energy from the Sun that was used by plants to make food. The Sun's energy, solar energy, can also be used directly to generate electricity. Solar energy can be harvested using solar panels which are made up of small units called photovoltaic cells. A photovoltaic cell works at the atomic level to convert light energy from the Sun directly into electrical energy. Materials used in photovoltaic cells have what is called a photoelectric effect. The photoelectric effect makes these materials able to absorb photons of light and release electrons. These released electrons are then captured, resulting in an electric current that can be used as electricity for power. Albert Einstein was granted a Nobel prize in physics in 1905 for describing the photoelectric effect.

Solar panels that convert the Sun's energy into electricity can be used in homes and are used to power spacecraft like satellites and the International Space Station. At present, solar power is relatively expensive, so it is less commonly used than energy from fossil fuels, flowing water, or wind. But as the technology for harvesting solar power improves, it will most likely become cheaper and more common.

Another way that solar energy can be used is called passive solar. Passive solar uses parts of a building itself to collect energy from the Sun. Materials that absorb heat can be selected for walls, floors, and other parts of buildings. Heat energy from the Sun is absorbed by these materials and released later, reducing the use of fossil fuels for heat. A greenhouse or large windows might be added to the south side of a building to allow for the collection of more solar energy. You may notice that when you get into a car that's been in the sun for a while, the air is warmer inside the car than outside. Your car can collect heat energy from the Sun, becoming a passive solar unit!

Solar panels

Photo Credit: US Air Force, Nellis AFB, Nevada/
photo by Airman 1st Class Nadine Y. Barclay

5.5 Summary

- Energy is converted from one form to another.

- Energy is conserved. This means that the total amount of energy does not change as energy is converted from one form to another.

- Usable energy can be lost when it is converted to unusable energy, such as light or heat, that is not captured and converted to another form of energy.

- Some of our energy comes from fossil fuels like oil, coal, and natural gas. These are forms of nonrenewable energy.

- Some of our energy is harvested from wind, flowing water, and the Sun. These are forms of renewable energy.

5.6 Some Things to Think About

- List some examples of energy being converted from one form to another.

- What do you think would happen if the total amount of energy on Earth were constantly changing, with energy being created and destroyed?

- Why do you think it's important for us to think about how we use energy?

- Do you think it could be important for communities to have different energy sources available? Why or why not?

- Do you think some energy sources might work well in one area but not in another? Why or why not?

- What advantages do you think renewable energy has over fossil fuels?

Chapter 6 Motion

6.1 Introduction

What is motion? We see objects moving every day. Cars move down the road. Planes move in the sky. We move as we work and play. Moving seems to be a very ordinary thing. However, figuring out exactly how objects move was a problem for early scientists. Although many people tried to figure it out, it took nearly 2000 years to finally understand!

According to history, the first person to study the science of motion was Aristotle. Aristotle was a Greek philosopher who was born in Stagira in 384 BCE. Aristotle thought that every moving thing was being constantly pushed from behind. He thought that the air in front of a moving ball was being separated as the ball moved and that the air behind would close up forcing the ball forward. He also thought that all moving objects moved because of this constant force. Based on these ideas, he thought that since he didn't feel the Earth moving, the Earth was sitting still. He also thought that the Sun and stars, because they changed places in the sky, were moving around the Earth. This belief system is based on a geocentric cosmos. *Geo* is the Greek word for "earth," and *centric* means "central." So geocentric means "earth centered." *Cosmos* is the Greek word for universe, so a geocentric cosmos means "earth-centered universe."

Today we know that the Earth moves around the Sun and that our solar system is heliocentric or "sun-centered." [*Helios* is the Greek word for "sun"].

Although Aristotle had many good ideas, he was wrong about motion. In the early 1600s Galileo finally showed with experiments that motions do not require a constant force to keep them going, and later Isaac Newton put these concepts into mathematical terms.

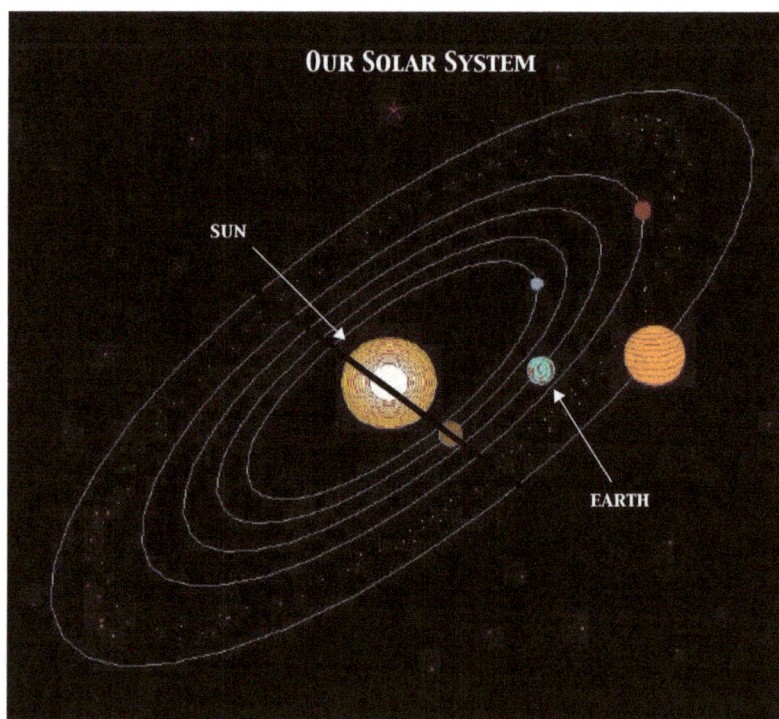

OUR SOLAR SYSTEM

SUN

EARTH

6.2 Inertia

If there is nothing pushing on a ball as it travels through the air, as Aristotle thought, how does it keep going? What keeps the ball moving at all? What Galileo discovered was that things always keep moving unless something stops them. This property is called inertia.

Simply put:

Inertia is the tendency of things to resist a change in motion.

This means that once something is moving, it will not stop, slow down, or change its direction unless something pushes on it. Aristotle had it completely backwards! It's not that forces keep things moving, but that forces *stop or change* the speed or direction of things that *are* moving!

Everything has inertia, no matter what it is — an atom, a rock, a baseball, or a car. Once the object gets going, it won't stop or even change direction unless something pushes on it.

6.3 Mass

The property that gives things inertia is called mass. Everything has mass, so everything has inertia. Because the force of gravity is constant everywhere on Earth, you can tell how much mass something has by weighing it.

For example, you can tell that a marble has less mass than a baseball because a marble weighs less than a baseball, and you can tell that a baseball has less mass than a bowling ball because a baseball weighs less than a bowling ball.

We know that once rolling, a marble would be much easier to stop than a baseball, and a baseball would be much easier to stop than a bowling ball. Can you imagine trying to bowl with marbles or trying to play marbles with bowling balls? The marbles don't have enough mass to knock down the pins, and the bowling balls have too much mass to roll with your thumb!

6.4 Friction

So, if inertia keeps things moving, what makes things stop? If you roll a ball, or push a toy truck, or take your foot off the gas pedal in your car, the ball, the truck, and the car all eventually stop moving. Why? If Galileo was right and inertia keeps things going, then wouldn't everything keep going all the time? Why do things stop?

What Galileo discovered is that objects keep going if no forces act on them. However, everyday objects like cars, bowling balls, and toy trucks almost always have forces acting on them. The force that makes objects stop is called friction. Friction is a force that tends to slow things down.

Usually friction is caused by things rubbing against each other. For example, if you slide a hockey puck on the street, the atoms in the hockey puck rub against the atoms in the street. The rubbing of these atoms against each other causes frictional force. The rougher the two surfaces, the more friction there is. If you could change the way the hockey puck contacts the street, you could reduce the friction. This is why hockey is not usually played on streets, but on ice. The ice is much smoother than the street, and the hockey puck slides much more easily. Galileo discovered that if he got rid of friction, things would keep on going and never stop. For example, if you could play hockey in space where there is no air and no friction, the puck would never stop! This is one of the most important discoveries in all of physics.

6.5 Momentum

Things are also harder to stop if they move fast. A baseball rolling slowly on the ground is easy to stop, but a baseball thrown by a pitcher is much harder to stop and requires a padded glove. So, there are two things that make something hard to stop: mass and speed.

In physics, the property that makes things hard to stop is called momentum. It has a precise mathematical definition:

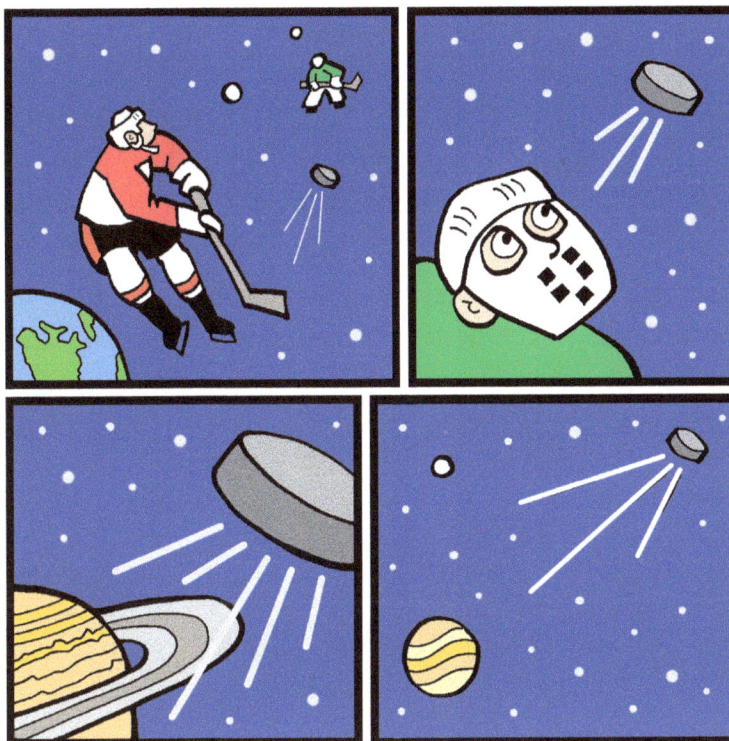

$$\text{momentum} = \text{mass} * (\text{speed} + \text{direction})$$
$$\mathbf{p} = \text{m} * \mathbf{v}$$

Where "**p**" stands for momentum, "m" for mass, and "**v**" for velocity. Velocity is speed + direction. We will learn more about velocity in the next chapter. (Note: "**p**" is the symbol for momentum because "**m**" is already being used for mass. "**p**" comes from the Latin word *petere* which means "to pursue or go towards.")

You can see from this equation that the more mass something has, the more momentum it will have. Also, the faster something travels (the more speed it has), the more momentum it will have. If something has lots of momentum, it will be harder to stop than something that has little momentum.

Momentum can change whenever there is a change in the mass of an object, the velocity, or both. But there is another factor that is important for changing momentum and that is time. Remember from Section 6.2 that forces stop or change the speed or direction of things that are already moving. If you apply a force for a longer period of time, you create a bigger change in speed or direction and thus a greater change in momentum.

For example, imagine that you are in a Soap Box Derby and your job is to get your team's go-kart moving fast enough that you can win. If you give the go-kart one strong push, it's not going to go very far. However, if you push the go-kart for a few yards with the same amount of force, the go-kart will travel farther. Why? The go-kart travels farther because you are changing velocity over a longer period of time, resulting in more momentum.

Knowing how time affects momentum can be useful. Understanding that momentum can increase as the length of time increases is great if you want to get a go-kart moving faster. But knowing that momentum can decrease as the length of time decreases can be handy in certain situations.

Momentum Math!

The relationship between **velocity**, **force**, and **time** can be described using mathematics and Newton's Laws. Newton figured out the following equation to show that **force**, **mass**, and a **change in velocity** over time were related:

Force = (mass) * (change in velocity ÷ time)

If we use some mathematics to rearrange this equation by multiplying both sides by time we get:

Force * time = change in (mass * velocity)

Because a change in **mass * velocity** equals **momentum** we get:

Force * time = momentum

This shows that the longer the time that a force is applied the greater the momentum will be.

Imagine you are sledding down a long hill and you discover you are going too fast to stop using only your feet. You can steer your sled into a fluffy pile of snow to the left or the side of a barn to the right. Which would you choose? You will hopefully choose the fluffy pile of snow because you know that the fluffy pile of snow will slow your momentum over a greater amount of time than the side of a barn, and by slowing your momentum more gradually you create less impact (force) as you stop, which is much easier on the body!

6.6 Summary

● **Inertia** is the tendency of objects to resist change in motion.

● **Friction** is a force that tends to slow objects down.

● **Momentum** depends on both the **mass** of an object and its **velocity** (speed + direction).

● Objects with large **mass** have more **momentum** than objects with small **mass**.

● Objects with a lot of speed have more **momentum** than objects with little speed.

● For a given applied force, **momentum** increases as length of **time** increases and decreases as **time** decreases.

6.7 Some Things to Think About

● How would you explain why people used to think the Sun moved around the Earth?

● Why do you think people were resistant to the idea of a heliocentric universe?

● What are some things that you think could push on a moving object to change its direction or slow it down?

● What is weight?

● Why do you think the discovery of the affects of friction is so important to physics?

● Think of a sport you have participated in or watched. How does momentum affect this sport?

Do you think having knowledge about the physics of momentum could help the athletes and the coaches perform better? How?

Chapter 7 Linear Motion

7.1 Introduction

How fast can you ride your bike? Can you race a car on your bike? Can you race a train? How fast does an airplane go? Does it go faster than a car? Does it go faster than a rocket? How can you measure how fast a bike, a car, or an airplane goes?

In the last chapter we looked at some general features of motion. We explored inertia and how an object will stay still or stay in motion because of inertia; how mass, momentum, and speed are related; and how friction will change or slow down the motion of an object. In this chapter we will take a closer look at a particular type of motion called linear motion.

Linear motion is, very simply, motion that occurs when any object travels in a straight line. In this chapter we will see that mathematics can be used to describe motion and that motion is defined by speed, acceleration, and velocity.

7.2 Speed

If you were to hop on your bike and cycle to the nearest grocery store for chocolate milk, how long would it take you? How long would it take to get to the same grocery store by car? Maybe you live in the remote wilderness and your only way to get to the store is by plane. How long would it take you to travel to the store by plane?

Whether you travel by bike, car, or plane, in each case you are going from one point to another. In other words, you are traveling a certain distance. If the distance is short enough, you could ride your bike. However, if the distance is far, you would probably want to take a car or even a plane.

The reason to choose a car or plane instead of a bike to travel a far distance is that a car or plane can go faster than a bike. In other words, a car or plane travels at a higher speed than a bike, and therefore a car or plane can travel a longer distance in less time than a bike.

Using mathematics we can calculate exactly how fast a car, plane, or bike can travel a certain distance. Speed is defined as the rate at which an object covers a given distance in a given amount of time. Recall that speed is defined mathematically by the following equation:

$$s = \frac{d}{t}$$

where "s" represents "speed," "d" represents "distance," and "t" represents "time."

To see how this equation allows us to calculate speed, let's imagine that you want to bicycle to the grocery store which is 16 km (10 miles) away. Because you're in really good shape, it only takes you 30 minutes to get to the store to buy chocolate milk. How fast did you go?

If we plug in the value of 16 km (10 miles) for distance and 0.5 hours for 30 minutes we get:

$$s = \frac{16 \text{ km (10 miles)}}{0.5 \text{ hrs}} = 32 \text{ km per hour (20 miles per hour)}$$

Suppose you don't know how long it might take you to ride your bike to the grocery store. But you do know how far it is to the store, and you know about how fast you can ride your bike. Can you figure out how long it will take? It's easy! All you have to do is rearrange the equation and *solve for time*.

Some Basic Math Rules, Terms, & Symbols

These symbols all mean divide: ___ , / , ÷

These formulas all mean that speed equals distance divided by time:

$$s = \frac{d}{t} \qquad s = d/t \qquad s = d \div t$$

These symbols all mean multiply: x, * , ·

These formulas all mean that momentum equals mass times velocity:

$$\mathbf{p} = m \times \mathbf{v}, \qquad \mathbf{p} = m * \mathbf{v}, \qquad \mathbf{p} = m \cdot \mathbf{v}$$

v   bold lower case "v" represents velocity

p   bold lower case "**p**" represents momentum

Δ   "delta" symbol—represents the difference between two quantities

| | "absolute value"—these upright lines mean that the result of a formula contained within them must be expressed as a positive number

variable - In an algebraic formula, a variable is a letter that represents a number that can change (vary).

1. In algebra when we perform the same operation to both sides of an equation, the two sides of the equation remain equal. We can add, subtract, multiply, or divide the same number (or variable) to both sides of the equation and the equation stays equal.

   For example, because $s = \dfrac{d}{t}$, then $s * 2 = \dfrac{d}{t} * 2$, or $s * x = \dfrac{d}{t} * x$, or $s * t = \dfrac{d}{t} * t$

2. In algebra if we multiply a variable by its inverse, we get 1.

   The inverse of a variable is its opposite, so the inverse of:

   $$\frac{s}{1} \text{ is } \frac{1}{s} \quad \text{and} \quad \frac{t}{1} \text{ is } \frac{1}{t} \quad \text{and} \quad \frac{d}{1} \text{ is } \frac{1}{d}$$

   If we multiply a variable by its inverse we get 1. For example:

   $$\frac{s}{1} * \frac{1}{s} = \frac{s}{s} = 1 \qquad \frac{t}{1} * \frac{1}{t} = \frac{t}{t} = 1 \qquad \frac{d}{1} * \frac{1}{d} = \frac{d}{d} = 1$$

   Any number divided by itself equals 1.

3. We can rearrange either side of an equation to help solve it. For example:

   $$s * t = t * \frac{d}{t} \quad \text{can be rearranged to} \quad s * t = \frac{t}{t} * d$$

   Because... $\dfrac{t}{t} = 1$ we get... $s * t = 1 * d$ so... $s * t = d$

Let's imagine that you want to go to the store in the next town because they have the most delicious chocolate milk. However, this store is 64 km (40 miles) away. If you can ride your bike at 32 km (20 miles) per hour, how long will it take to get to the store?

Here's how we can use some basic math (algebra) to *solve for time.* First we multiply both sides of the equation by "t" (time):

$$s = \frac{d}{t} \quad \text{(speed = distance divided by time)}$$

$$s*t = t * \frac{d}{t} \quad \text{(both sides multiplied by time)}$$

Because $t * \dfrac{1}{t} = 1$, the "t"s on the right hand side cancel each other out:

$$s*t = \cancel{t} * \frac{d}{\cancel{t}}$$

Results: $s*t = d$ (speed multiplied by time = distance)

Now, if we divide both sides by "s" (speed) we get:

$$\frac{s*t}{s} = \frac{d}{s}$$

Because $s * \dfrac{1}{s} = 1$, the "s"s on the left hand side cancel each other:

$$\frac{\cancel{s}*t}{\cancel{s}} = \frac{d}{s}$$

Results: $t = \dfrac{d}{s}$ (time = distance divided by speed)

Using this equation we can figure out how long it will take to go 64 km (40 mi.) at 32 km (20 mi.) per hour. Plugging in the values for "d" and "s" we get:

$$t = \frac{64 \text{ km}}{32 \text{ km per hour}} \quad \left(\text{or } t = \frac{40 \text{ miles}}{20 \text{ miles per hour}} \right)$$

Solution: $t = 2$ hours

Because the trip will take two hours, you might want to grab an extra chocolate milk for the ride home! Or, if you don't have four hours for a trip to the grocery store and back, you could go by car.

You Do it!

1. Your car will go 80 km per hour (50 miles per hour). How long will it take you to get to the grocery store and back if the store is 80 km (50 miles) away?

2. Which bicycle route will get you there faster?
   Route A: 30 km with an average speed of 15 km per hour,
   or
   Route B: 30 km with an average speed of 10 km per hour

See solution steps below.

Solution Steps

1. **Calculate for time**

   80 km/h (50 mph) = speed = s
   80 km (50 miles) = distance = d
   We are looking for "time" = t

   To solve:

   $$\text{speed} = \frac{\text{distance}}{\text{time}} \qquad s = \frac{d}{t}$$

   Multiply both sides of the equation by "t":
   $$s*t = \frac{d*t}{t}$$

   The "t"s cancel:
   $$s*t = \frac{d*\cancel{t}}{\cancel{t}} \qquad\qquad s*t = d$$

   Divide both sides by "s":
   $$\frac{\cancel{s}*t}{\cancel{s}} = \frac{d}{s}$$

   $$\boxed{t = \frac{d}{s}}$$ The "s"s cancelled so this is my equation!

   Plug in variables:
   $$t = \frac{80 \text{ km}}{80 \text{ km/h}} = 1 \text{ hour}$$

2. **Route A or B?** (See next page for solution steps.)

Solution Steps (continued)

2. **Route A or B?**

I'll use the equation for time.

To solve for Route A:

$d = 30$ km

$s = 15$ km/h

$t = \dfrac{d}{s} = \dfrac{30 \text{ km}}{15 \text{ km/h}} = \boxed{2 \text{ hr}}$ Route A is faster!

To solve for Route B:

$d = 30$ km

$s = 10$ km/h

$t = \dfrac{d}{s} = \dfrac{30 \text{ km}}{10 \text{ km/h}} = 3 \text{ hr}$

In the previous example, we assumed a constant speed. In other words, we assumed that there are no stop signs or slow traffic or any other reasons to go faster or slower on our trip to the grocery store.

However, on most roadways there are stop signs and traffic congestion that will change the speed you are traveling. If you are riding a bicycle, there may be hills that cause you to go slower on the way up and faster on the way down.

When we use the total distance traveled divided by the total time it takes, we are actually calculating the average speed. The speed of the car or bicycle will change depending on whether or not there are other cars or bicycles on the road, turns to make, and traffic signs to obey. By using the total time and total distance, we don't see the changes in speed from one section of the trip to the next. However, if you have a speedometer, you can watch how fast or slow you go in certain sections of the trip and calculate how much time it would take you to travel only those sections. Knowing how long it might take to travel certain sections might change the route you take.

For example, suppose your house is in the middle of the block and in the middle of a steep hill. To get to the store you first have to get out of your neighborhood, and you can choose to either go to the north and up the hill or to the south and down the hill. Even if the distance is longer, you might decide to go south, riding a little farther but going downhill so you can ride faster. Your average speed would be different for each route because the speed you travel over a particular section would be different.

7.3 Velocity

In the last example we added "north" and "south" to our discussion of speed. North, south, east, and west are all directions. A direction describes where you are headed relative to your current position. If you say, "I am going north to the bowling alley," you are describing the direction of the bowling alley relative to your current position.

In physics, speed plus direction is called velocity. If you say you are going 25 km per hour, you are describing your speed. However, if you say you are going 25 km per hour to the north, you are describing velocity.

Although speed and velocity look similar, in physics they are actually quite different. Both speed and velocity have distinct but different mathematical meanings. Speed is a scalar quantity and velocity is a vector quantity. (See next page for definitions.)

Scalars and Vectors

Mathematically, speed is defined to be a scalar quantity. In math, the word scalar simply means an amount or magnitude. For example, if a car is traveling at 40 km per hour, this describes the *amount* of speed at which the car is traveling from one point to another. A scalar is simply a number that represents a value like speed, temperature, weight, or height. In an equation, a scalar is often written as a lower case letter. For example: s, d, t

Mathematically, velocity is defined to be a vector quantity. In math, a vector has both magnitude (amount) and direction (up, down, right, left, north, south, etc.). A vector is often written as a lower case letter in a **bold** font, such as "**v**" for velocity.

We can think of a vector as an arrow. The length of the arrow represents the magnitude, and the orientation of the arrow represents its direction.

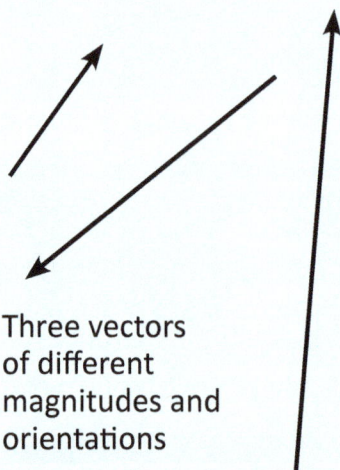

Three vectors
of different
magnitudes and
orientations

Speed (a scalar) is the magnitude (numerical value) for velocity (a vector). The length of the arrow represents the numerical value of the magnitude.

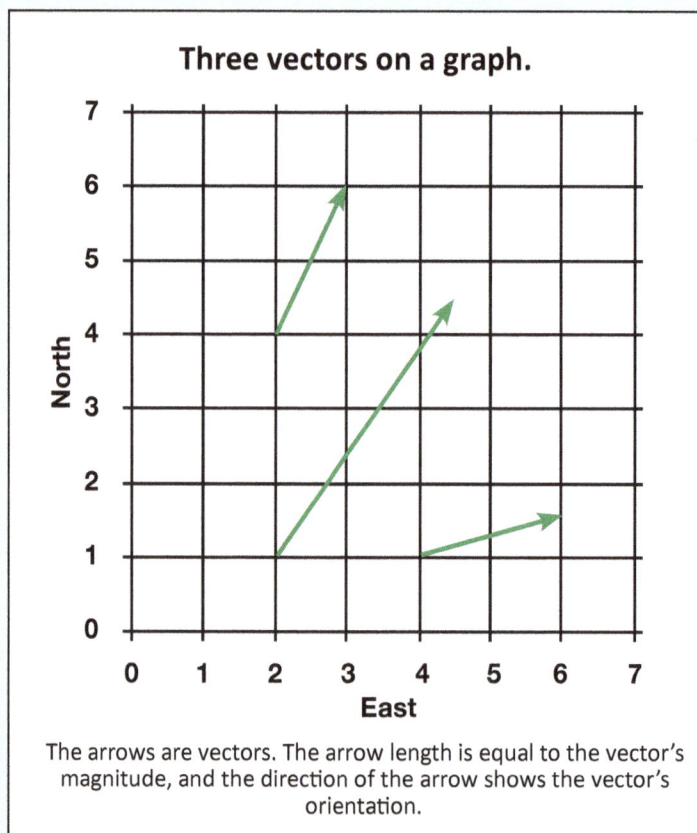

Three vectors on a graph.

The arrows are vectors. The arrow length is equal to the vector's magnitude, and the direction of the arrow shows the vector's orientation.

Technically speaking, speed is the rate at which an object moves, and velocity is the rate at which an object changes its position.

On a linear path the speed and velocity have the same magnitude (or amount). If a horse going north trots 13 km (8 miles) in one hour from point A to point B, the horse's speed is:

$$s = \frac{d}{t} = \frac{13\,km}{1\,hr} = 13\ km\ per\ hour$$

And the velocity is:

$$\mathbf{v} = \frac{d}{t} = \frac{13\,km}{1\,hr} = 13\ km\ per\ hour\ north$$

What happens if a horse is trotting on a circular path? Because velocity takes direction into account, speed and velocity are different.

Imagine the horse is on a 1 km (.6 mile) track and trots around the track 13 times, traveling 13 km in one hour and returning to the starting line. The speed, or rate, at which the horse trotted is:

$$s = \frac{d}{t} = \frac{13\,km}{1\,hr} = 13\ km\ per\ hour$$

However, since the horse stopped and started at the same place, there was no change in its position and the velocity would be zero! (See the math on the next page.)

Strange but True Velocity Math!

In the example of the horse trotting around the track, we can use an illustration to see how the velocity is equal to zero. If we divide the track up into different sections and add the velocities of each section, it looks like this:

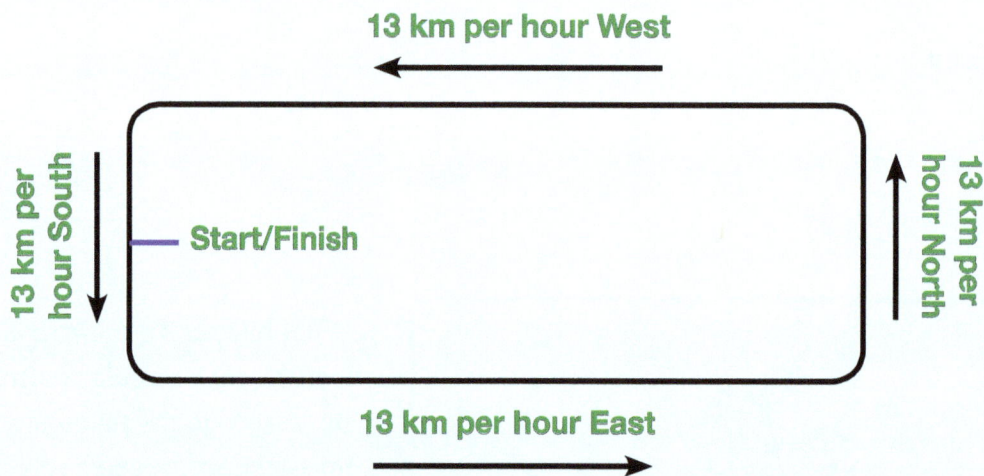

13 km per hour West

13 km per hour South

13 km per hour North

— **Start/Finish**

13 km per hour East

The total velocity of the horse is:

Total v_T = 13 km/h north - 13 km/h south + 13 km/h east - 13 km/h west

$$v_T = v_n - v_s + v_e - v_w$$

Because velocity is a vector, the directions cancel each other and we see that:

$$v = 0$$

In this example the velocity of the horse is zero even though it maintained a constant speed of 13 km per hour. When the horse returned to its starting position, there was no change in its position from when it began trotting, so the total velocity is zero.

7.4 Acceleration

Sometimes velocity changes. If you are riding your bike and come to a big hill, you might slow down as you climb up. When you go over the top and then have a long descent, you will most likely go faster. In both cases your velocity will change because both your direction and speed change as you slow down or speed up and go up or down the hill.

You have probably felt the effects of going faster or slowing down on a bike, in a car, or in a plane. If you are racing to the finish line on your bike, you might spin your legs faster to win! When you spin faster, you can feel your body jerk a little as you suddenly speed up. If you are in a car and the light turns yellow, you might press on the brake to slow down to a full stop. As you quickly push on the brake, you can feel your body move forward. In both cases you can feel the moment when you suddenly change velocity.

We change the velocity of something by changing its speed, its direction, or both its speed and direction. When velocity changes over a given time it is called acceleration. Acceleration can be represented by the equation:

$$\mathbf{a} = \frac{\Delta \mathbf{v}}{\Delta t}$$

where "**a**" represents acceleration, "Δ (delta) **v**" represents the change in velocity and "Δ (delta) t" represents the change in time. Acceleration equals the change in velocity divided by the change in time.

The small delta symbol (Δ) is a mathematical symbol that represents the difference between two quantities. In this equation "Δ**v**" stands for the difference between the final velocity and the initial velocity, and "Δt" stands for the difference between the final time and the initial time. When we expand the equation by plugging in the variables for the initial and final velocities and times, we get:

$$a = \frac{v_f - v_i}{\left| t_f - t_i \right|}$$

Where "v_f" is the final velocity, "v_i" is the initial velocity, "t_f" is the final time, and "t_i" is the initial time. Also, for acceleration, time is always a positive number. "Δt" represents the "change in time." When the result of a calculation must be a positive number, or absolute value, the calculation is placed between two upright lines. In the acceleration formula, the change in time is written as $\left| t_f - t_i \right|$ to show that the result is expressed as a positive number.

Acceleration MATH!

Example 1:

Calculate the acceleration of a bowling ball going from zero to 2 meters per second in 2 seconds.

$$a = \frac{v_f - v_i}{\left| t_f - t_i \right|} = \frac{2 \text{ m/sec.} - 0 \text{ m/sec.}}{\left| 2 \text{ sec.} - 0 \text{ sec.} \right|} = \frac{2 \text{ m/sec.}}{2 \text{ sec.}} = 1 \text{ m/second}^2$$

Example 2:

Calculate the acceleration of a projectile slowing down from 50 meters per second to 10 meters per second in 10 seconds.

$$a = \frac{v_f - v_i}{\left| t_f - t_i \right|} = \frac{10 \text{ m/sec.} - 50 \text{ m/sec.}}{\left| 10 \text{ sec.} - 0 \text{ sec.} \right|} = \frac{-40 \text{ m/sec.}}{10 \text{ sec.}} = -4 \text{ m/second}^2$$

You Do the Acceleration MATH!

Physics Math Problem 1

1. Calculate acceleration where $v_f = 10$ m/sec., $v_i = 25$ m/sec., $t_f = 5$ seconds, $t_i = 0$

2. Calculate acceleration where $v_f = 40$ m/sec., $v_i = 10$ m/sec., $t_f = 10$ seconds, $t_i = 0$

(See Appendix for solution.)

7.5 A Note About Math

You can see in this chapter that you can describe linear motion exactly with mathematics. Learning mathematical terms like scalar and vector, symbols like delta, and algebra rules can make linear motion both clearer and more complicated, depending on how much math you understand.

As we discussed in the last chapter, mathematics is an essential tool for physics. Because physical actions like linear motion can be described exactly using math, scientists have been able to launch rockets into space, put people on the Moon, and explore the possibility of traveling to other worlds.

The best way to learn the math is to practice. By understanding and solving the problems in this book and making up your own problems over and over again, you can master physics and math!

7.6 Summary

- Linear motion is the motion of an object in a straight line.

- Linear motion is described by three quantities — speed, velocity (speed + direction), and acceleration.

- Speed is defined by the distance traveled divided by time: s = d÷t

- Velocity is defined by the distance traveled in a particular direction divided by time: **v** = d÷t

- Acceleration is defined by the change in speed (or velocity) divided by the change in time: $a = \dfrac{\Delta v}{\Delta t}$

- Mathematics is an essential tool for doing physics.

7.7 Some Things to Think About

- What are some examples of linear motion?

- List some experiments or projects a scientist might be working on that would require the calculation of speed.

- Do you think calculating speed could be important to athletes? Why or why not?

- In your own words, define velocity and speed.

- How would you describe acceleration?

- Make up your own physics math problems and share them with a friend.

Chapter 8 Non-Linear (Curved) Motion

8.1 Introduction

In the last chapter we looked at how linear motion is defined by speed, velocity, and acceleration. We also looked at the equations for calculating speed and acceleration and how these equations work.

But what happens when motion is not linear? How can you calculate the speed of an object whose path curves as it moves? What happens when a ball rolls? Does the speed, velocity, or acceleration change? What happens when the wheels of a car rotate or when a cannon ball goes up and then down again?

In this chapter we will take a look at non-linear motion (curved motion), or the motion of objects that don't follow a straight line. One type of non-linear, or curved, motion includes the motion of any object (cannon ball, baseball, bullet, etc.) that follows a curved path. Another type of non-linear motion includes the motion of objects that rotate, or move in a circular pattern, like a car wheel, bike gear, or airplane propeller.

8.2 Projectile Motion

When you throw a basketball through a hoop, or a football down a field, or launch a pumpkin from a catapult, you are projecting a projectile. Projecting means "throwing, casting, or moving an object forward," and a projectile is the object that is being thrown, cast, or moved forward.

The motion of a projectile is described by its trajectory. A trajectory is the path the projectile takes at each instant in time once it is launched, thrown, or cast forward. When a ball or shot put is thrown in the air, the trajectory is non-linear (curved). In both of these situations, the projectile is launched into motion and is pulled toward Earth by gravity, resulting in a curved path.

Recall from the last chapter that linear motion can be described by speed, velocity, and acceleration. For an object traveling with linear motion, speed and velocity are often used interchangeably. As the object remains traveling in a straight line, the velocity need not change because the direction does not change. However, for non-linear (curved) motion the direction does change, and velocity and speed can no longer be interchanged.

The velocity for non-linear motion can be broken down (divided) into two components — a vertical component and a horizontal component. The vertical component describes the speed of the projectile directly perpendicular to the Earth's surface, and the horizontal component describes the speed of the projectile directly parallel to the Earth's surface. The total velocity is the sum of the vertical and the horizontal components of the velocity.

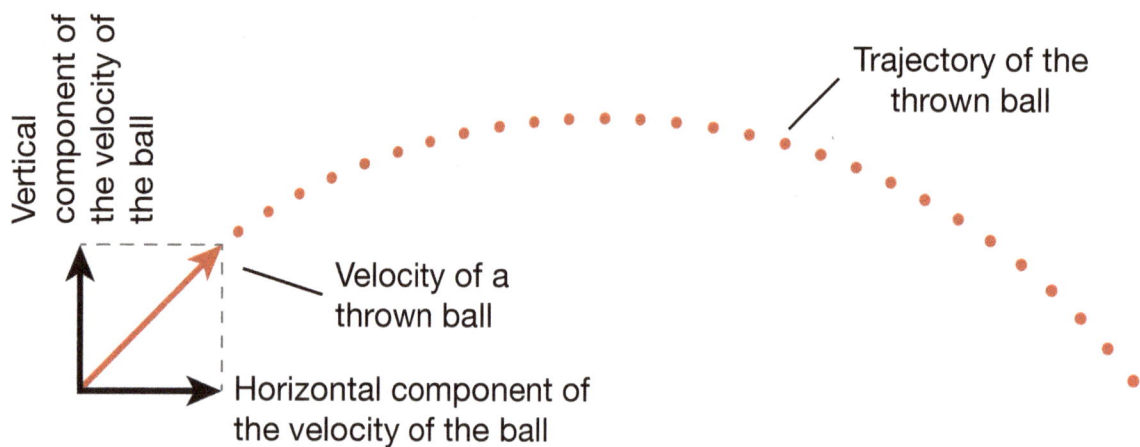

Vertical component of the velocity of the ball

Trajectory of the thrown ball

Velocity of a thrown ball

Horizontal component of the velocity of the ball

As the ball travels completing the trajectory, the velocity of the vertical component changes as gravity pulls the ball toward Earth. However, the horizontal component stays the same unless there is air resistance or some other force pushing the ball backwards. The total velocity changes as the ball travels through the arc.

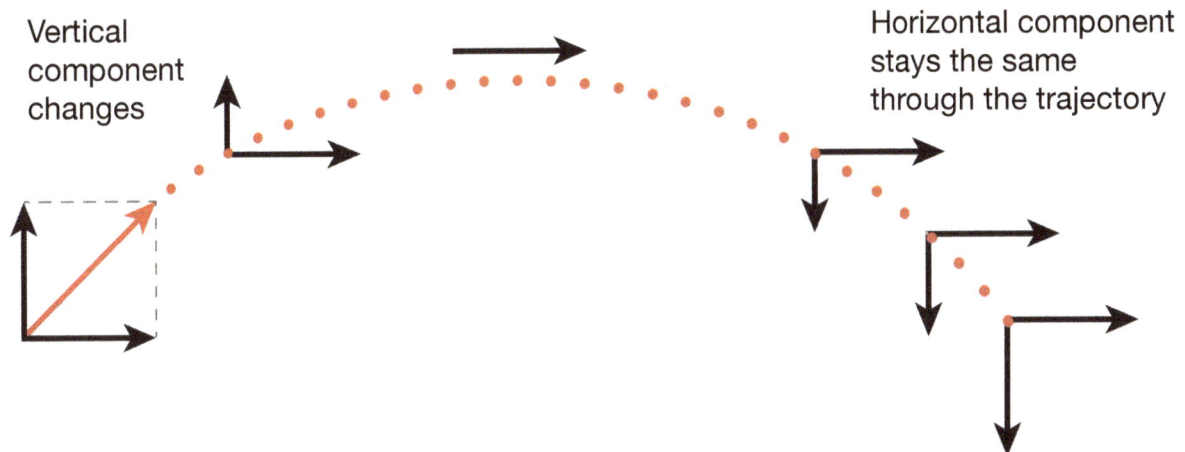

Vertical component changes

Horizontal component stays the same through the trajectory

The distance a projectile will travel depends on the initial speed and the angle at which it is projected (direction). If the ball is thrown at too high an angle, it goes up and back down without going very far. If the ball is thrown at too shallow an angle, it reaches the ground before it has had a chance to travel very far. However, if the ball is thrown at a 45 degree angle, it has just the right combination of speed and direction to travel the farthest!

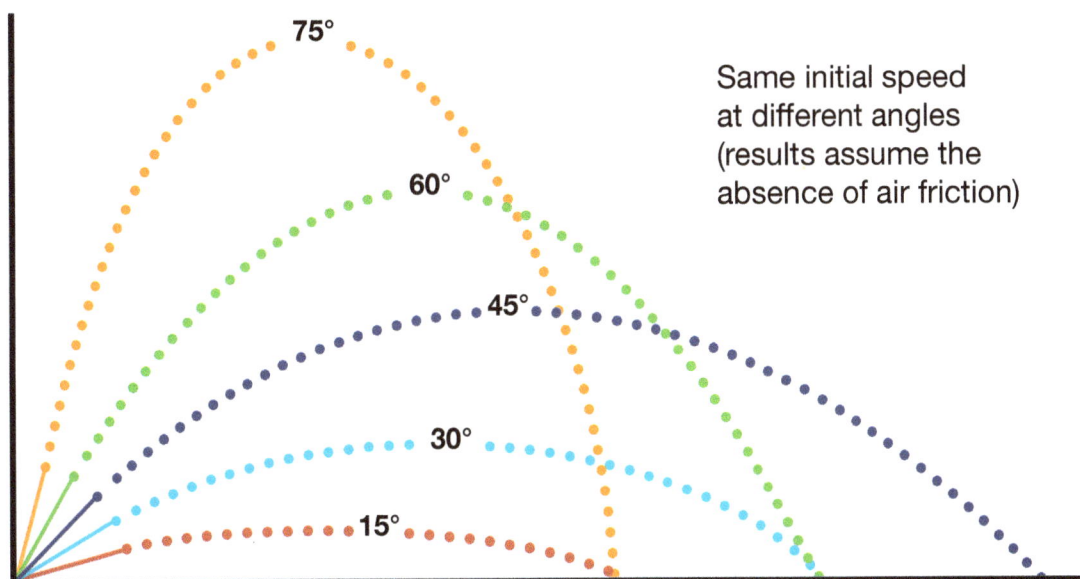

75°

60°

45°

30°

15°

Same initial speed at different angles (results assume the absence of air friction)

How can knowing about projectiles, angles, and speed help in the Punkin Chunkin competition? By understanding the physics, you can know that a pumpkin launched as a projectile will follow a curved trajectory. And knowing that a curved trajectory has both vertical and horizontal velocities, you can pick a path that predicts the distance the pumpkin will travel. You can also pick a launching mechanism that produces a high initial velocity that can be directed at a 45 degree angle. By applying your knowledge of physics, you can win the competition!

8.3 Circular Motion

Have you ever played on a merry-go-round? If you have, you probably noticed that when you push the merry-go-round, you run in a circle, and when you hop on and sit on the merry-go-round, it spins you in a circle. Moving in a circle is a special type of non-linear, or curved, motion called circular motion.

Circular motion has two different types of speed — tangential speed and rotational speed.

Tangential speed

Perpendicular line tangent to circle

Circle center

Circle

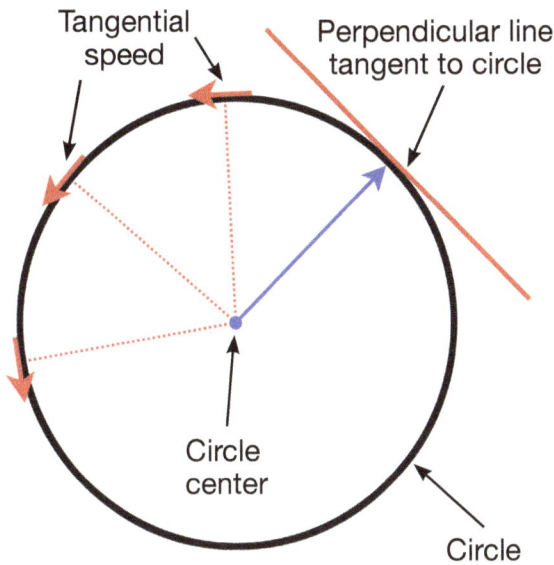

Tangential speed is the distance you travel along the path of the circle (tangential to the circle) over a given time. A tangent is a perpendicular line that intersects any point on a circle. Every point on the circle has a tangent. One way to think about tangential speed is to imagine an arrow traveling around the outer edge of the circle. It has tangential speed as it travels around the circle.

You may also have noticed that if you sit on the outside edge of the merry-go-round, you go faster than if you sit in the center. When you are farther from the center of a circle, the distance you travel is greater than when you are closer to the center of the circle so you have to go faster to make one complete revolution. Recall that speed is simply the distance traveled divided by time, so it makes sense that the tangential speed is greater farther from the center of the circle than it is closer to the center.

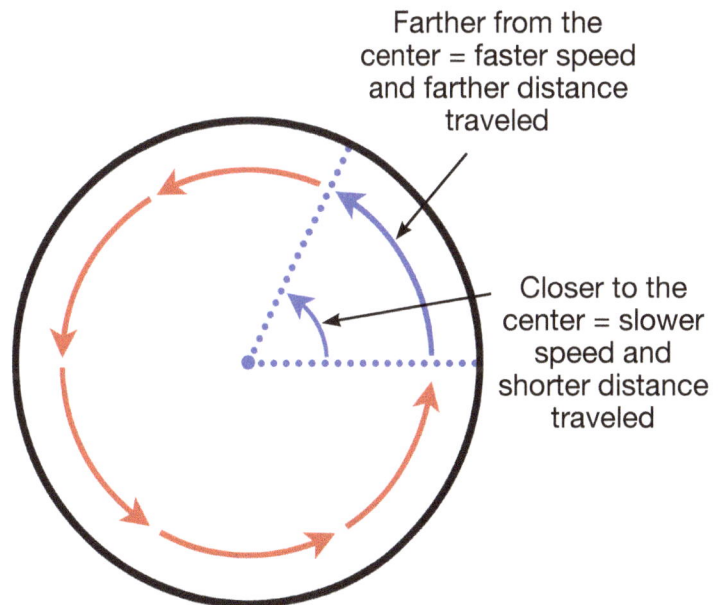

Farther from the center = faster speed and farther distance traveled

Closer to the center = slower speed and shorter distance traveled

You Do It!

Physics Math Problem 3

Is there any tangential speed at the center point of the circle?

(See Appendix for solution.)

Rotational speed (also called angular speed) is determined by how many rotations a circle makes around its central axis in a certain amount of time. Rotational speed is the same both farther away and closer to the center of the circle. Rotational speed is often expressed as revolutions (rotations) per minute (RPM). Some cars have a gauge on the dashboard that measures the RPM of the engine. This gauge tells the driver how fast the engine is rotating the crankshaft of the motor.

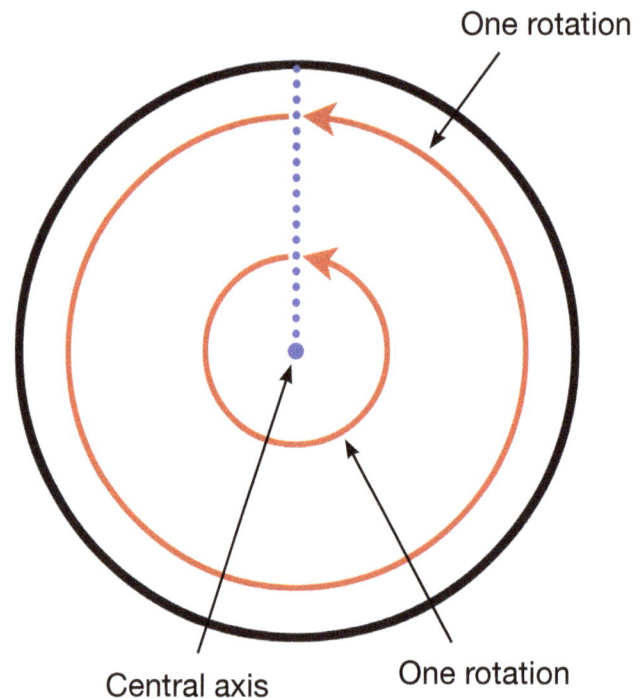

One rotation

Central axis   One rotation

Some bicyclists use a meter called a cadence meter that measures how fast they are rotating the pedals. Cadence meters measure the RPMs of the pedals and help cyclists determine which gear to use to get the most speed for the least effort.

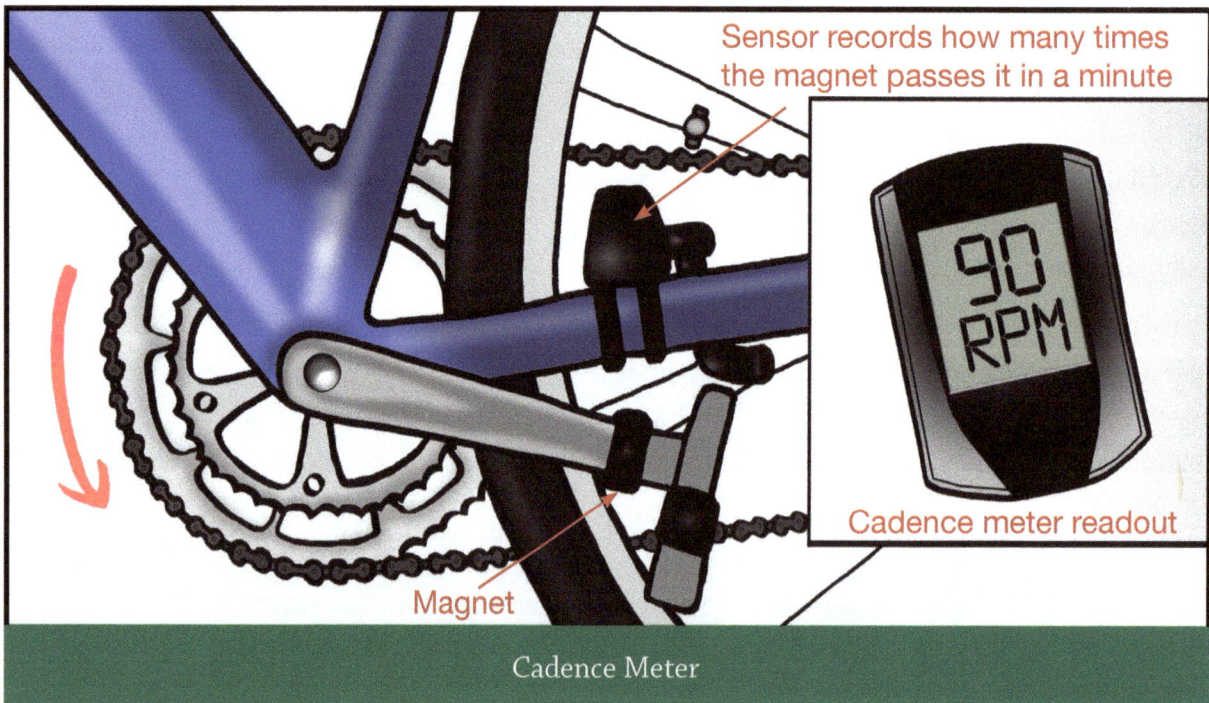

Sensor records how many times the magnet passes it in a minute

90 RPM

Cadence meter readout

Magnet

Cadence Meter

You Do It!

Physics Math Problem 4

Place your finger on a rotating disk, such as a spinning bicycle wheel, other wheel, or a toy spinning top that has a flat upper surface, and let your finger move with the disk. Observe the difference between rotational speed and tangential speed. Place your finger on the outside of the rotating object and observe how far your finger travels in one revolution. Now place your finger on the inside of the rotating object and observe again how far your finger moves in one revolution. In both cases your finger traveled one rotation, so the rotational speed is the same.

1. Which position (closer or farther out) has more tangential speed?

2. Which position (closer or farther out) has more rotational speed?

3. What happens when you place your finger directly in the center?

(See Appendix for solution.)

8.4 Summary

● Non-linear motion (curved motion) is any motion that does not travel in a straight line.

● A projectile is any object that is being thrown, cast, or launched.

● A trajectory is the path a projectile travels.

● The distance a projectile will travel depends on the angle and the initial speed at which it is projected.

● Circular motion is non-linear motion (curved motion) that travels in a circle.

● Circular motion is defined by both tangential speed and rotational speed.

8.5 Some Things to Think About

● List examples of non-linear motion.

● In your own words, describe the two components of the velocity for non-linear motion.

● The Punkin Chunkin competition is an example of a time when you would want to know about projectiles, angles, and speed. What are some other events or experiments when this knowledge would be essential?

● Describe the two different types of speed for circular motion.
How do you think you could measure each type of speed?

Chapter 9 Chemical Energy

9.1 Introduction

What is chemical energy? Simply put, chemical energy is the energy that is released by chemical reactions.

For example, we know that if we add vinegar to baking soda, a chemical reaction occurs. We can observe the chemical reaction as the properties of both the baking soda and the vinegar change, and bubbles of carbon dioxide gas are released.

The energy given off in chemical reactions can be used to do work. Recall that work is done when a force acts on an object and causes it to move or change shape. Chemical energy can be converted to other types of energy and can do work. For example, imagine what would happen if we put baking soda and vinegar in a jug and then put a cork on top. Pretty soon, the cork would POP off the jug! The gas produced by the chemical reaction of the vinegar and baking soda would push up on the cork—applying a force. The force produced can become strong enough to pop off the cork. In this case, the chemical reaction produced gas that did work on the cork. In the next section we'll take a closer look at how the chemical reaction between baking soda and vinegar can do work.

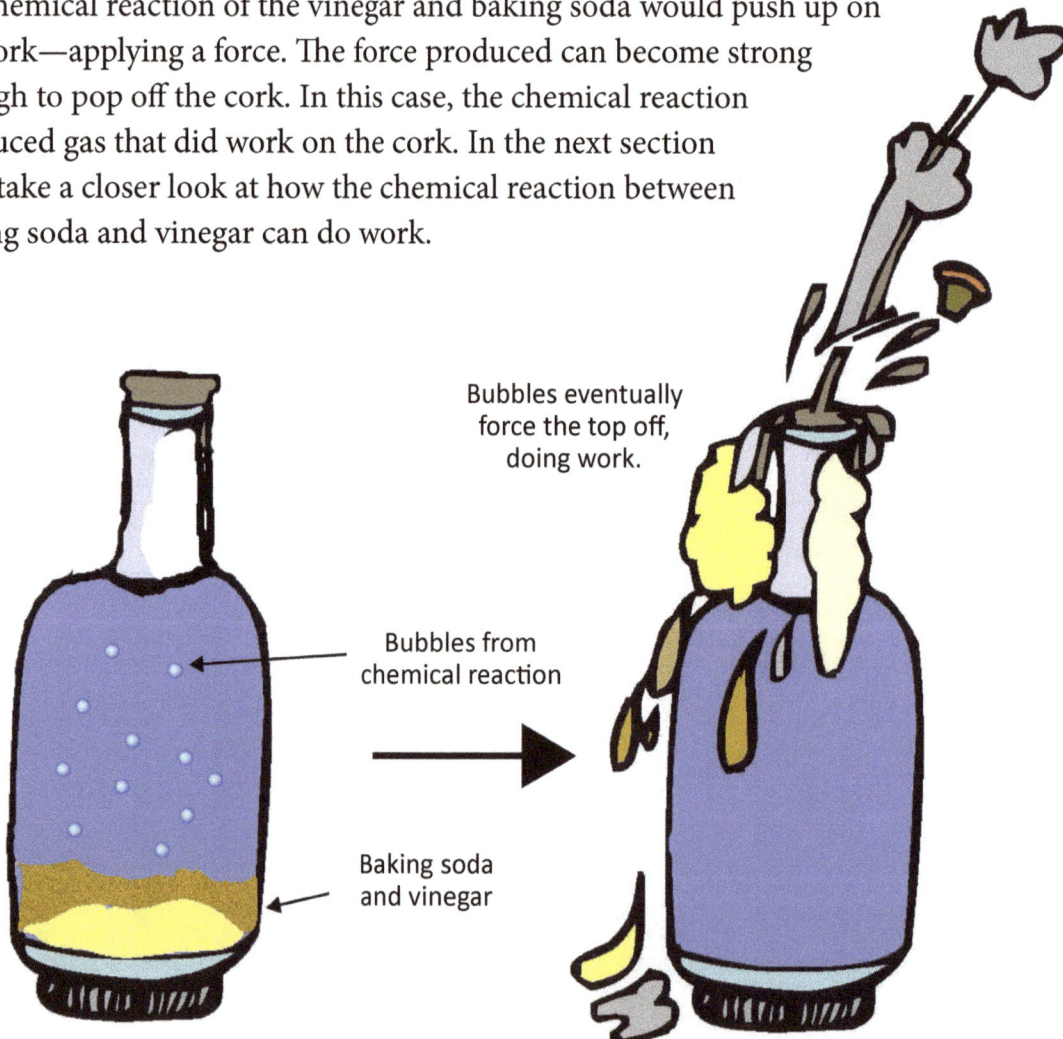

Bubbles eventually force the top off, doing work.

Bubbles from chemical reaction

Baking soda and vinegar

9.2 Gas Laws

First, let's take a look at the chemical reaction between baking soda and vinegar. This chemical reaction is an acid-base exchange reaction.

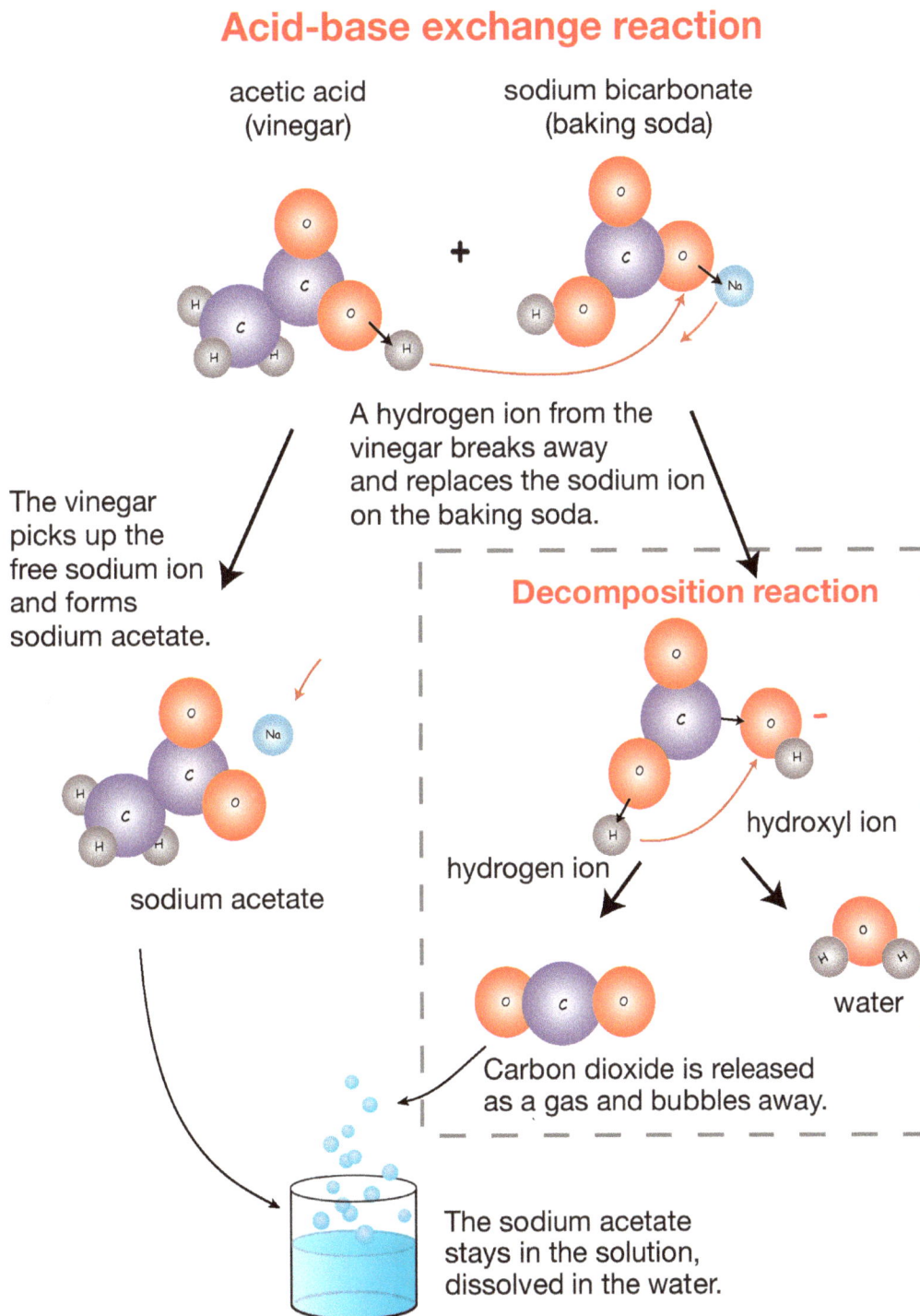

Acid-base exchange reaction

acetic acid
(vinegar)

sodium bicarbonate
(baking soda)

A hydrogen ion from the vinegar breaks away and replaces the sodium ion on the baking soda.

The vinegar picks up the free sodium ion and forms sodium acetate.

Decomposition reaction

sodium acetate

hydroxyl ion

hydrogen ion

water

Carbon dioxide is released as a gas and bubbles away.

The sodium acetate stays in the solution, dissolved in the water.

One of the first things we notice when we look at this reaction is that some compounds are solid, some are liquid, and some are gas. This becomes important when we start to explore how a chemical reaction can do work. Baking soda is a white powder and is added to the reaction as a solid. Both vinegar and water are liquids, and carbon dioxide is a gas. You can watch the carbon dioxide gas bubble off as it escapes from the liquid salt water (sodium acetate in solution) into the atmosphere. The sodium acetate stays dissolved in solution, but if all the water is evaporated, the sodium acetate will be left behind as a solid.

We can write the overall chemical equation as:

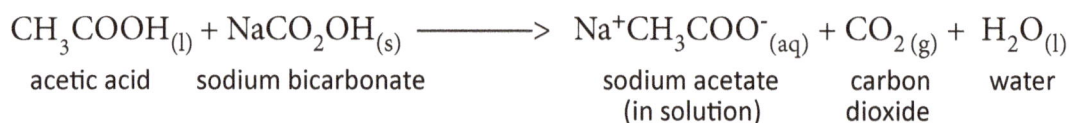

$$CH_3COOH_{(l)} + NaCO_2OH_{(s)} \longrightarrow Na^+CH_3COO^-_{(aq)} + CO_{2\,(g)} + H_2O_{(l)}$$

acetic acid   sodium bicarbonate   sodium acetate   carbon   water
(in solution)   dioxide

The subscripts (l), (s), (aq), and (g) describe the physical state of matter of the compounds: l = liquid, s = solid, aq = aqueous solution, and g = gas.

Although solids and liquids can expand and contract somewhat, gases can expand and contract significantly. We can say that gases are highly compressible, while liquids and solids are almost incompressible. In this reaction, the ability to expand into large spaces and contract into very small spaces gives the carbon dioxide gas the ability to do work. How does this happen?

To understand how a gas can do work, we need to take a closer look at the typical properties of gases. In general, a gas has four measurable properties that describe it: *pressure, volume, temperature, and the quantity of the gas.*

Pressure

Gases have higher molecular energy than either solids or liquids. This high amount of energy makes gas molecules move very fast, bouncing around in all directions. If a gas is confined in a bottle or container, the gas molecules will push against the walls of the container, creating pressure.

Mathematically, pressure is defined as force per unit area.

$$P = \frac{F}{A}$$ where P=pressure, F=force, and A=area

To find the pressure created by a gas, we need to find out how much force is produced by the moving molecules.

Because gas molecules are too small and are moving too quickly to measure with a ruler and a scale, we need to look at the relationships between pressure, volume, temperature, and the quantity of gas.

There are a series of equations collectively called the Gas Laws that describe the relationships between these four quantities—pressure, volume, temperature, and quantity of gas. It is useful to describe in general how these relationships work.

Pressure and Volume

One gas law states that pressure is inversely proportional to the volume of the container the gas occupies. In other words, for the same number of molecules of gas, if the volume of the container increases, the pressure goes down. If the volume of the container decreases, the pressure goes up. You can see this in action when you pop the air-filled bubbles in bubble wrap. When you press on a bubble, you are effectively reducing the volume of the container and increasing the pressure. If you press hard enough, you create enough pressure to pop the bubble.

The mathematical equation that relates pressure and volume looks like this:*

$$P \propto \frac{1}{V}$$

Where P = pressure (inside the bubble)
        V = volume (inside the bubble)
        $\propto$ = mathematical symbol meaning "proportional to"

Gas molecules in a container

Same number of gas molecules in a larger container creates lower pressure

Same number of gas molecules in a smaller container creates higher pressure

Pressure and Volume

For the same number of gas molecules: Changing the size of the container changes the size of the surface area the total number of molecules come in contact with. This changes the pressure (force per unit area).

*The number of gas molecules is constant.

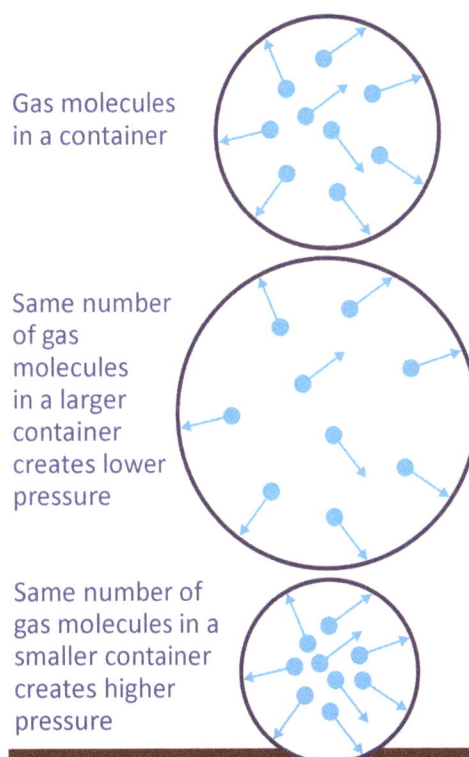

Temperature and Volume

What happens if you put an air-filled balloon in the freezer? The temperature inside the freezer is much lower than the temperature in the room. When you take the balloon out of the freezer, you can see that it is smaller. The volume inside the balloon goes down as temperature goes down. As the air in the balloon warms up to room temperature, the balloon will once again expand to a larger volume as the gas in it expands.

Raising the temperature increases how fast the gas molecules are moving, which increases the pressure. Lowering the temperature decreases how fast the gas molecules are moving, which decreases the pressure. Because a balloon is made of flexible material, as the gas warms up and the molecules have more energy to push on the sides of the balloon, the balloon will stretch. This causes the volume of the balloon to get larger as the pressure stretches the balloon. When the gas is cooled, the molecules have less energy to push on the sides of a balloon, and the volume of the balloon decreases as the pressure on the sides decreases. We can summarize this by saying that the volume of a gas is proportional to the temperature.

Mathematically, the equation that relates volume to temperature looks like this:

$$V \propto T$$

Where V = volume (inside the balloon)

T = temperature (inside the balloon)

$\propto$ = mathematical symbol meaning "proportional to"

Volume and Quantity of Gas

When you are blowing up a balloon, you can see that up to a certain point, the balloon continues to expand with every breath. Each time you exhale, you are forcing more air molecules into the balloon, and because the balloon is made of flexible material, it will stretch and the volume will become larger. The greater the number of air molecules that are forced into it, the more the balloon will expand. The greater the number of molecules of gas, the larger the volume of the container if the material it is made of is not rigid. So, for a given pressure, volume is proportional to the quantity of gas (assuming a stretchy container). Mathematically, the equation that expresses this is:

$$V \propto n$$

Where V = volume (inside the balloon)

n = number of moles of gas molecules (inside the balloon)

$\propto$ = "proportional to"

The Ideal Gas Law

Now that we know how pressure relates to volume and volume relates to temperature and the quantity of gas relates to volume, we can combine all of these into one equation called the Ideal Gas Law. Using the Ideal Gas Law we can finally show how the gas from a chemical reaction can pop the cork off a bottle.

The Ideal Gas Law is:

$$PV = nRT$$

Where P = pressure

V = volume

n = number of moles

T = temperature

R is the Ideal Gas Law Constant = 0.082057 L•atm/K•mol:

where L= liter, atm = atmospheric pressure,

K = Kelvin, and mol = number of moles)

The gas constant "R" is needed to combine the previous equations.

Using the Ideal Gas Law equation we can find out how much pressure is exerted by the carbon dioxide gas. Let's assume that we use 1 mole of acetic acid and 1 mole of sodium bicarbonate in the vinegar/baking soda acid-base exchange reaction equation.

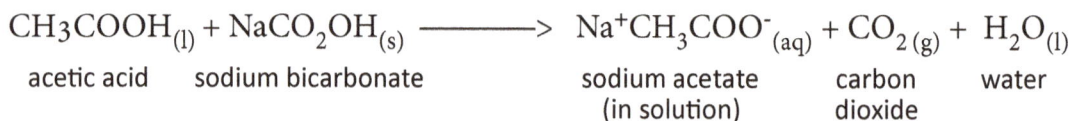

$$CH_3COOH_{(l)} + NaCO_2OH_{(s)} \longrightarrow Na^+CH_3COO^-_{(aq)} + CO_{2(g)} + H_2O_{(l)}$$

acetic acid   sodium bicarbonate   sodium acetate   carbon   water
                                            (in solution)   dioxide

Because our equation is balanced, we can see that 1 mole of acetic acid reacts with 1 mole of sodium bicarbonate to produce 1 mole of sodium acetate, 1 mole of carbon dioxide, and 1 mole of water. Let's also assume the reaction is taking place at a room temperature of 25°C or 298 K (kelvin) [see the following inset].

Kelvin and Other Temperatures

Scientists use kelvin (K) as the unit of measure for temperature in place of Celsius and Fahrenheit. The formula for converting Celsius to kelvin is K=°C + 273. Kelvin measurements do not include the word "degree" or its symbol.

	Celsius (°C)	Kelvin (K)	Fahrenheit (°F)
Surface of the Sun	5,600	5,900	10,100
Boiling Point of Water	100	373	212
Body Temperature	37	310	98.6
Hot Day	30	303	86
Room Temperature	20	293	68
Cool Day	10	283	50
Freezing Point of Water	0	273	32

We now have all the information we need to complete the calculation and can make a list:

- V = 1 liter (volume of the bottle); Volume is both the size of the space in the container and the space the gas fills.
- n = 1 mole of carbon dioxide gas
- R = 0.082057 L•atm/K•mol
- T = 298 K

Next, we can plug these values into the equation. First, though, let's rearrange the equation so P is on one side of the equation and everything else is on the other side:

$$P = \frac{nRT}{V}$$

Plugging in our values we get:

$$P = \frac{(1\ mol)}{1\ L}\left(\frac{0.082057\ L \cdot atm}{K \cdot mol}\right)298\ K$$

All of the units cancel except for atm:

$$P = \frac{(1\ mol)}{1\ L}\left(\frac{0.082057\ L \cdot atm}{K \cdot mol}\right)298\ K$$

$$P = (0.082057 \cdot 298) = 24.4\ atm$$

To get some perspective on this number, consider that normal atmospheric pressure is 1 atm which is equivalent to 14.7 pounds per square inch at sea level. This means that the pressure exerted by the carbon dioxide in the 1 liter container is more than 24 times atmospheric pressure. Another way to think about it is to consider that 24 atm is about 100 times the amount of pressure in a bike tire. This is a lot of pressure! With that much pressure pushing on the cork, it's easy to see how the cork will *pop off!*

9.3 Stored Chemical Energy

Before chemical energy is used or converted into other forms of energy, it is stored. Stored chemical energy is called chemical potential energy. It is called potential energy because it has the "potential" to do work.

In the previous example we found out how the gas formed in a chemical reaction can pop the cork off a bottle. When atoms and molecules form or break bonds, chemical energy is released. In the reaction between acetic acid and sodium bicarbonate, chemical energy is released and new molecules, such as carbon dioxide, are formed. Carbon dioxide gas has fast moving molecules that bounce against the walls of the container and the cork, creating pressure. The carbon dioxide gas does work as it forces the cork to pop off.

There are different ways to store chemical energy. Many fuels are stored chemical energy. Petroleum, for example, is a form of stored chemical energy. Crude petroleum (unrefined oil) is made primarily of hydrocarbons, with smaller amounts of heterocompounds which are hydrocarbon molecules that contain sulfur, nitrogen, and other atoms in addition to hydrogen and carbon.

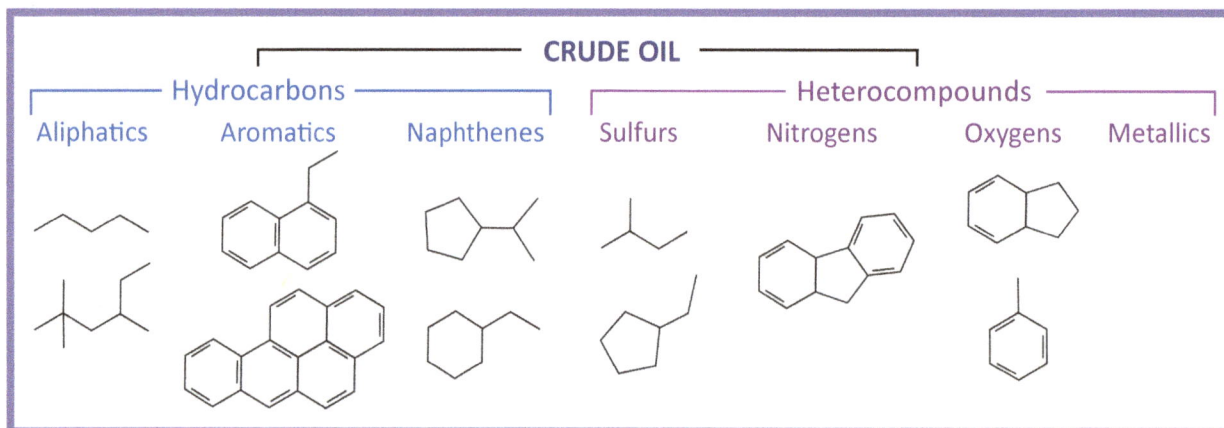

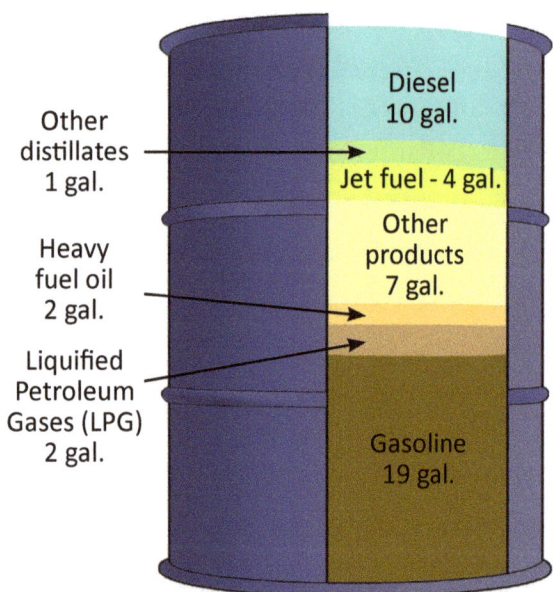

A variety of different fuels can be made from crude oil and include diesel fuel, gasoline, jet fuel, and other products.

When petroleum products like gasoline are ignited in the presence of air, they combust, or "burn," and are converted into carbon dioxide, water, and heat, which can be used to do work.

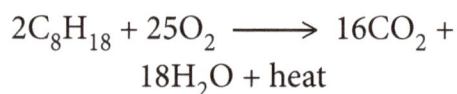

$$2C_8H_{18} + 25O_2 \longrightarrow 16CO_2 + 18H_2O + heat$$

The engines in gasoline fueled cars are called internal combustion engines because the combustion of gasoline occurs within the engine. A car does work by using the energy produced by chemical reactions.

Petroleum products from crude oil

A barrel of crude oil holds 42 gallons. During the refining process other substances are added, causing the refined products to measure as much as a total of 48 gallons.

(Quantities of refined products in this illustration are approximate.)

Spark plug

Spark plug wire

Intake valve

Exhaust valve

Fuel mixture intake

Exhaust

Piston

Cylinder

Connecting rod

Crankshaft

Some basic parts of an internal combustion engine

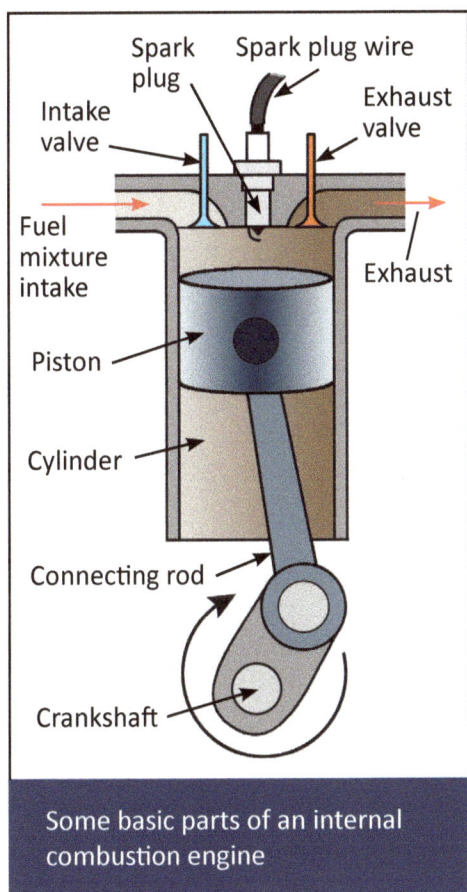

Combustion occurs within an enclosed space called the cylinder. A piston moves up and down within the cylinder and fits tightly enough that gas can't escape between the piston and the sides of the cylinder. Valves at the top of the cylinder open and close to control the flow of fuel into and exhaust out of the cylinder. A connecting rod attaches the piston to the crankshaft, and the motion of the end of the connecting rod rotating around the crankshaft moves the piston up and down.

Most gasoline powered cars have four-stroke engines with each movement of the piston up or down considered to be a stroke. The four strokes are called intake, compression, power, and exhaust. At the beginning of the intake stroke, gasoline and air are mixed together in the carburetor and the fuel mixture is sent through a chamber to the cylinder. The intake valve is open, allowing the fuel mixture to enter the cylinder above the piston. The piston is moving downward, creating low air pressure within the cylinder. The higher atmospheric pressure outside the cylinder pushes the fuel-air mixture through the open valve into the low pressure space in the cylinder.

The intake valve closes as the compression stroke begins and the piston starts to move upward, compressing the fuel-air mixture which is now trapped in the cylinder above the piston. Reducing the volume of the fuel-air mixture increases the pressure within the cylinder and heats the fuel, providing more energy for combustion.

When the piston has reached the top of the cylinder in the compression stroke, the power stroke begins. Both valves are still closed to seal the gases in the cylinder. An electric current is sent to the spark plug located at the top of the cylinder. The spark produced by the spark plug makes the fuel explode, producing carbon dioxide in the combustion reaction. When the gas expands, it forces the piston down, providing energy to turn the crankshaft and ultimately the wheels of the car.

The last stroke in the cycle is the exhaust stroke. As the piston begins once again to move upward, the exhaust valve opens. The piston forces the by-products of combustion out of the cylinder. Once the piston has reached the top of the stroke, the exhaust valve will close. As the piston begins to move downward, the intake stroke will begin and the cycle will repeat. Most gasoline powered cars have 4, 6, or 8 cylinders that fire at different times and are attached at different angles on the crankshaft. This provides continuous energy to the crankshaft to keep it turning smoothly.

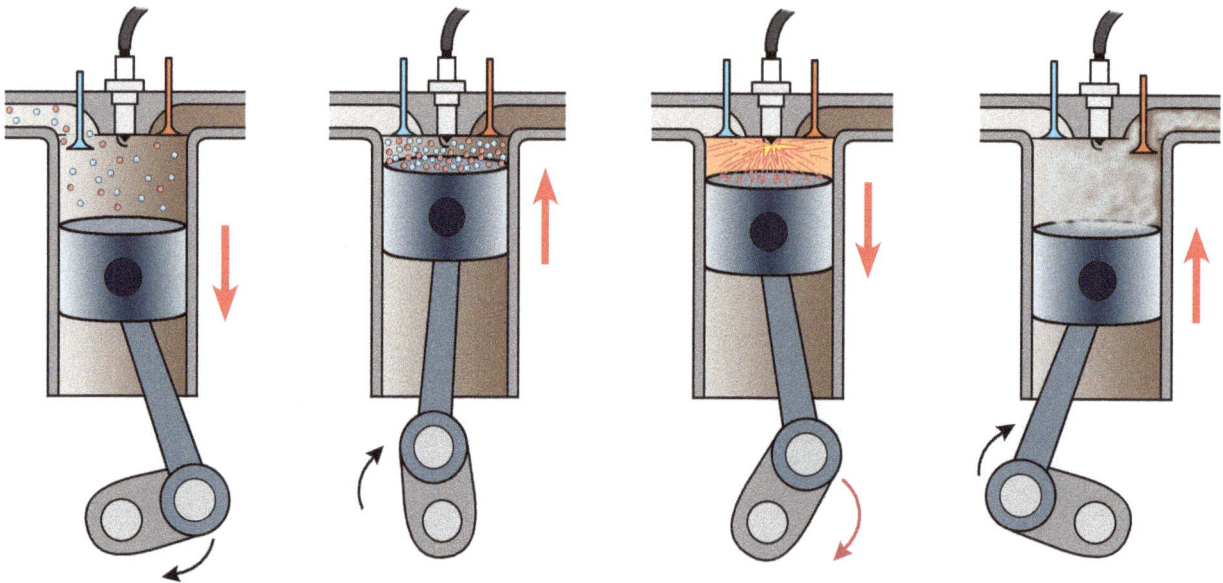

Intake Stroke	Compression Stroke	Power Stroke	Exhaust Stroke
The intake valve is open and the piston is moving downward, creating a low pressure area. The air-fuel mixture is entering the cylinder.	Both valves are closed and the piston is moving upward, compressing the air-fuel mixture.	Both valves are closed. A spark ignites the hot, compressed gases, forcing the piston down, turning the crankshaft, and powering the car.	Exhaust valve is open, intake valve closed. The piston is moving up, forcing the by-products of combustion out of the cylinder.

One problem with internal combustion engines is that they cause pollution. The carbon dioxide produced as a by-product of internal combustion engines goes into the atmosphere as a greenhouse gas. It is thought that by increasing the amount of greenhouse gases in the atmosphere, cars are contributing to climate change. Also, the gasoline used to power an internal combustion engine doesn't completely combust, and unburned hydrocarbons are emitted into the air as particles of soot. This particulate matter can cause respiratory problems when breathed in. Also released into the atmosphere are carbon monoxide gas and oxides of nitrogen and sulfur. Sulfur oxides contribute to acid rain which can damage plants and buildings, and they are irritating to the eyes and respiratory tract. Nitrogen oxides can be harmful to plants and animals, and carbon monoxide is a toxic gas.

9.4 Stored Chemical Energy in Food

Chemical potential energy is also found in food. All of the food we eat is a form of stored chemical energy, and some foods, like potatoes, have lots of stored chemical energy. In the same way that cars convert fuel into mechanical energy through combustion, human bodies convert starches into energy in a much slower and, overall, more efficient combustion reaction.

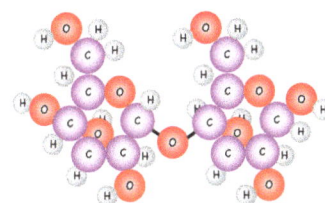

Starch

Glucose

Fructose

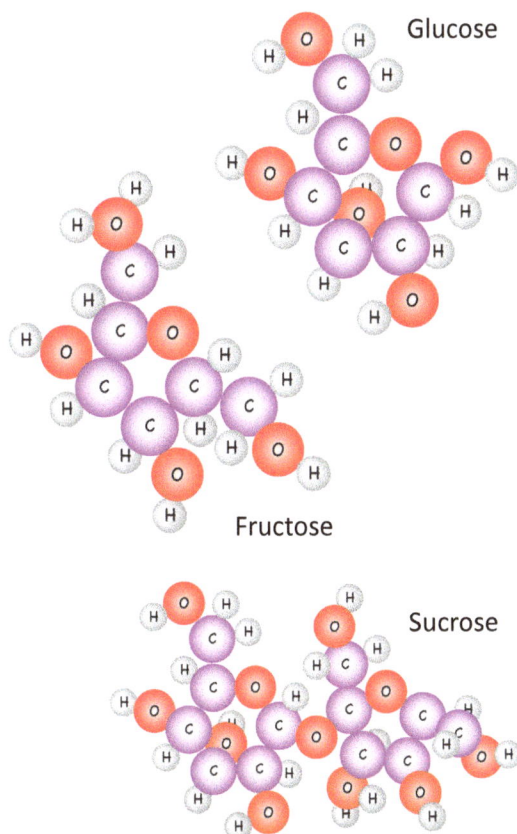

Energy molecules used by the body for fuel

Sucrose

To convert glucose molecules into energy inside the body requires a series of several chemical reactions taking place inside the cells. The first series of chemical reactions used to convert carbohydrates into energy is called glycolysis. Glycolysis occurs inside the cytosol of cells. The cytosol is the aqueous part of the cell outside the nucleus. It has organelles suspended in it. Glycolysis converts glucose into a molecule called pyruvate which is used in a later reaction called the citric acid cycle where ATP and NADH are made. ATP and NADH are the primary "fuel" molecules the body uses for energy.

Fats are also used to produce energy in the body. Fats do not go through glycolysis but are instead broken down and enter the citric acid cycle directly. Fats and fatty acids are long chain molecules of hydrogen and carbon. Proteins can also be used for fuel but require more energy

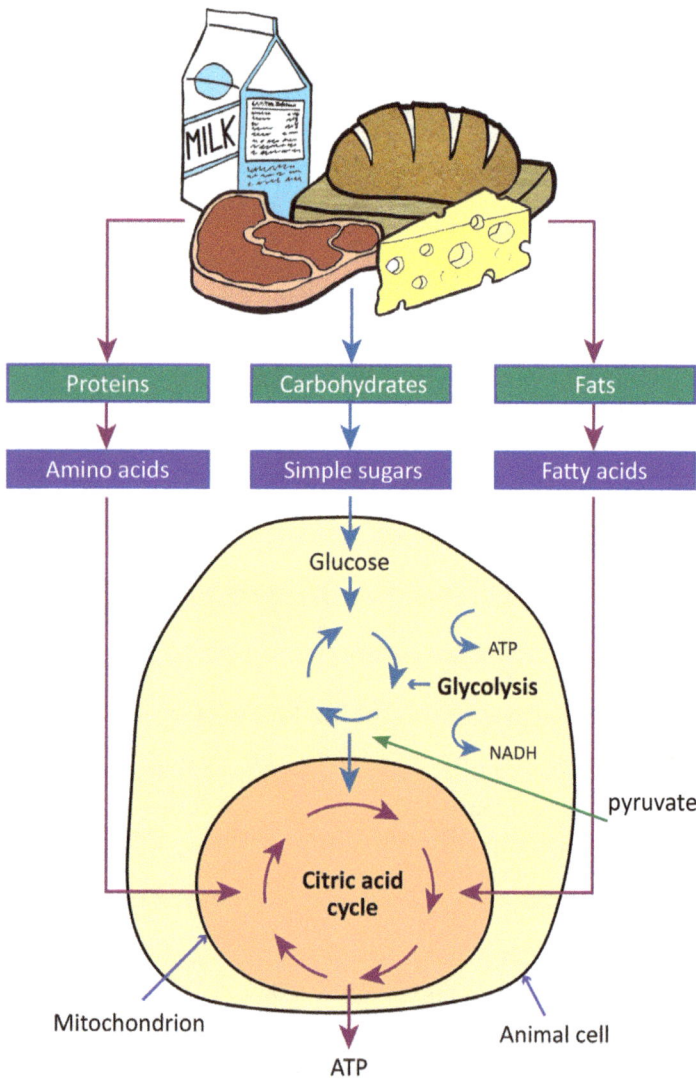

Proteins

Carbohydrates

Fats

Amino acids

Simple sugars

Fatty acids

Glucose

ATP

← **Glycolysis**

NADH

pyruvate

Citric acid cycle

Mitochondrion

Animal cell

ATP

Cells in the body break down food using glycolysis and the citric acid cycle. The energy molecules ATP and NADH are produced and fuel the body.

to break down than either carbohydrates or fats and are therefore not utilized by the body as a primary energy source. Instead, proteins are used by the body to grow and repair cells.

The body burns carbohydrates and fats through oxidation. Oxidation as used by the body is similar in some ways to how a car burns gasoline. Just like oxygen is used to convert petroleum products into energy through chemical reactions, oxygen is used to convert both carbohydrates and fats into carbon dioxide and water, creating energy in the form of ATP that can be used by the muscles. ATP is a molecule that provides energy needed by a cell for chemical reactions to take place. In this way, chemical potential energy in carbohydrates and fats is converted to mechanical energy in the muscles.

9.5 Stored Chemical Energy in Batteries

Batteries represent another type of stored chemical energy. The Italian scientist Alessandro Volta (1745-1827) constructed the first battery, which was made of alternating layers of silver and zinc disks separated by pieces of cloth soaked in salt water. This type of battery is called an electrochemical or voltaic cell.

A voltaic, or battery, cell creates a difference in stored potential energy through chemical reactions. In a voltaic cell, alternating layers of metal disks and electrolyte soaked cloth are stacked on top of each other. (An electrolyte is a liquid that contains ions.) A specific set of chemical reactions called oxidation-reduction reactions, or redox reactions, occur at the interface between each alternating metal disk and electrolyte soaked cloth. Briefly, these reactions either release or uptake electrons. This causes electrons to build up on one side of the voltaic cell and be depleted on the other side. The result is a difference in stored potential energy, and when a wire or other conducting path connects the two sides, electrons flow from one end to the other, generating electric energy.

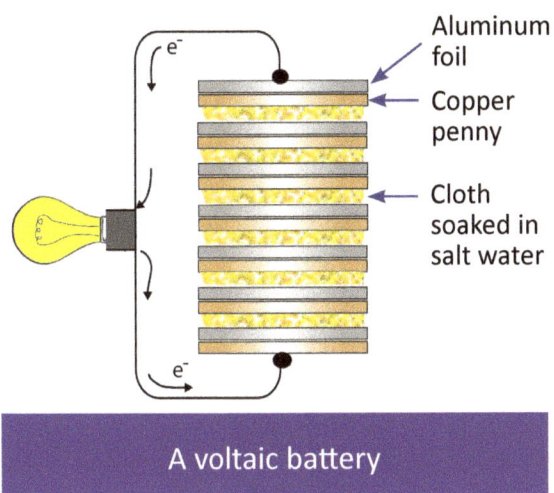

Aluminum foil

Copper penny

Cloth soaked in salt water

A voltaic battery

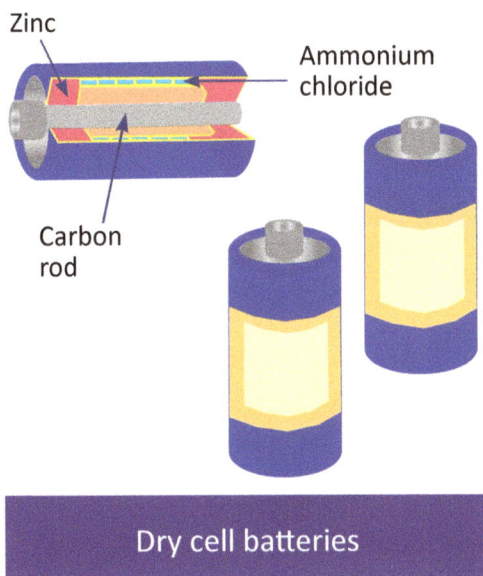

Zinc

Ammonium chloride

Carbon rod

Dry cell batteries

One type of voltaic cell is called a dry cell which is the kind of battery we put in flashlights, toy cars, and cell phones. A dry cell gets its name because the chemicals it uses are not liquids but pastes. The chemicals in the pastes are chosen to store a lot of energy so the battery lasts a long time. They are also designed to avoid giving off gases or creating any dangerous by-products.

9.6 Summary

● Chemical energy comes from chemical reactions between atoms and molecules.

● A gas has four measurable properties that describe it: pressure, volume, temperature, and the quantity of the gas.

● The equations that show how the four measurable properties of gas interact with each other are called Gas Laws.

● Stored chemical energy is called chemical potential energy because it has the "potential" to do work when it is converted into other forms of energy, such as electrical energy or mechanical energy.

● The food we eat is a form of stored chemical energy.

● Batteries generate electrical energy from chemical reactions.

9.7 Some Things to Think About

● How would you define chemical energy? What can it do? Give an example.

● What are the four measurable properties of a gas?

● What are the Gas Laws and how are they used?

● Why is stored chemical energy also called chemical potential energy?

● How is stored chemical energy used in an internal combustion engine?

● How does the body get energy from food?

● Why do you think dry cell batteries stop working?

● What products have dry cell batteries made possible?

Chapter 10 Electrostatics

10.1 Introduction

What happens when you rub your stocking feet across the carpet and touch a doorknob?

If you live in a very dry climate like New Mexico, you are likely to feel a slight shock as your finger makes contact with the doorknob. This shock is the result of electric charges being built up on your feet, traveling up to your finger, and then being quickly transferred from your body to the doorknob, causing a slight electrical shock. This is commonly called static electricity. The word static comes from the Latin word *stare* which means "to stand." Static electricity is caused by standing (not moving) electric charges. In physics, static electricity is also called electrostatic charge.

People commonly refer to electric charges and their movement as electricity. The word electricity is a broad term used to describe a wide variety of different phenomena including everything from lightning bolts to batteries. We can generate "electricity" with batteries, wind, or water, and we use "electricity" when we plug in our toasters or turn on our kitchen light. If we drive an electric car, we can experience a car powered entirely by "electricity." But to understand the science behind "electricity," it's important to take a closer look at electric charges and also how they move. The areas of physics that study electric charges are: electrostatics (electric charges at rest) and electrodynamics (moving electric charges). Electrostatics is presented in this chapter, and electrodynamics is presented in Chapter 11.

Electrostatics, the study of electric charges at rest, includes understanding how electric charges work, the forces between electric charges, and their behavior in different types of objects or materials. While exploring the subject of electrostatics, we will take a closer look at how electric charges are collected by objects like your stocking feet and body.

10.2 Electric Charge

An atom has protons, neutrons, and electrons. Protons are positively charged, electrons negatively charged, and neutrons are neutral and have no charge. All electrical phenomena begin with electric charge at the atomic level.

Charges with the same sign will repel each other, and charges with opposite signs will attract each other. Because electrons are negatively charged, they will want to be as far away from each other as possible. However, because protons and electrons have opposite charges, they will want to be as close as possible to each other. This attraction between oppositely charged particles is what holds atoms together. Attraction is also what holds molecules together. All chemical bonding is the result of attractive electrical forces between atoms.

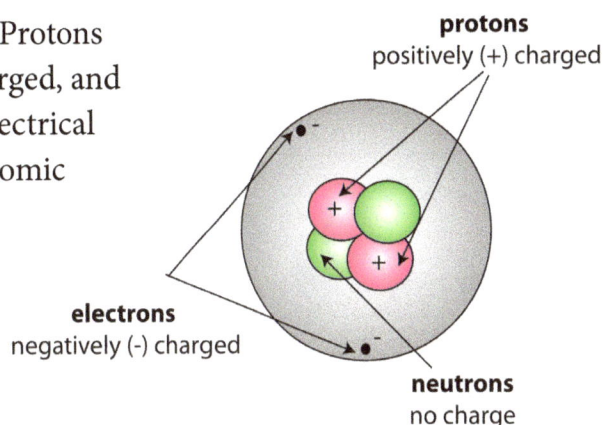

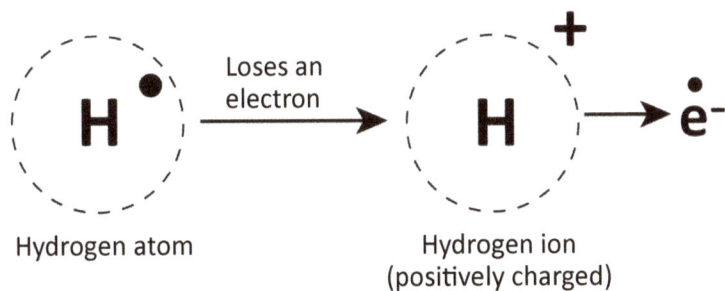

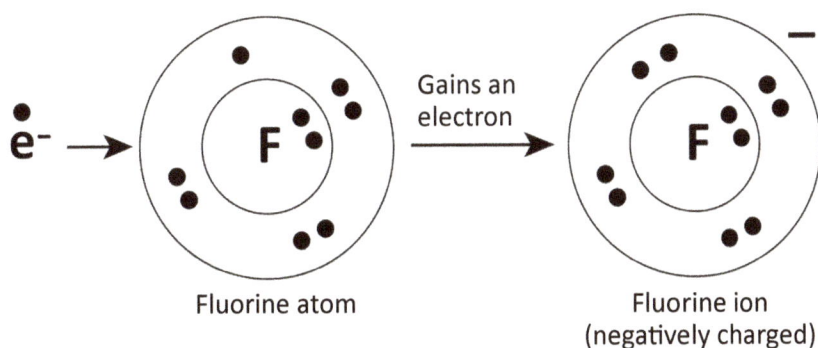

We also know from chemistry that an atom with equal numbers of protons and electrons will have no charge; however, if an electron is added or removed, the atom will become charged. A charged atom is called an ion. A positive ion is an atom that has a net positive charge because it has lost one or more electrons, and a negative ion is an atom that has a net negative charge because it has gained one or more electrons.

All materials are made of atoms. This means that everything around us is made of protons, neutrons, and electrons. This book, the one you are reading right now, is made of protons, neutrons,

Helium Atom

protons
positively (+) charged

electrons
negatively (-) charged

neutrons
no charge

Hydrogen atom

Loses an electron

Hydrogen ion
(positively charged)

Fluorine atom

Gains an electron

Fluorine ion
(negatively charged)

and electrons. But notice that as you hold the book in your hands, you don't get an electric shock, nor can you use the book to power a toy car or heat a toaster oven. The book has no charge because the atoms that make up the book have an equal number of protons and electrons. In other words, the book is not "charged." Objects ordinarily have equal numbers of protons and electrons and are not charged.

However, if you live in a dry enough climate, you might be able to transfer electrons to the surface of an object using friction. If you take a balloon and rub it in your hair and then pull the balloon away slightly, you may notice that your hair now clings to the balloon. Electrons from your hair are transferred to the balloon, making it charged. If you transfer enough electrons, you can even stick the balloon on a wall! As electrons from an object are added or removed, a slight imbalance occurs, and this imbalance will cause the object to become charged.

10.3 Van de Graaff Generator

The amount of electric charge you can generate with balloons and hair in a dry climate is pretty small, but with a Van de Graaff Generator you can generate much more electrostatic charge.

The Van de Graaff generator, a device used to generate electrostatic charge, was invented in 1929 by American physicist Robert J. Van de Graaff (1901-1967 CE). To generate electrostatic charge, this

device has a belt that moves in a loop around two rollers. Because the belt and the rollers are made of different materials, electrostatic charges of opposite signs build up on the belt. At the top of the device is a hollow, spherical, metal dome. As the electrons leave, a net positive charge from the atomic nuclei in the atoms remains on the surface . These positive charges create an electric field that radiates in all directions from the surface of the metal dome.

Van de Graaff generators are often used in science museums and classrooms to demonstrate electrostatic charge. If you place your hand on the metal sphere, electric charges will travel over your body to your hair. Your charged hairs will repel each other, making your hair stand up on end. Definitely a bad hair day!

10.4 Electric Force

What keeps like charges away from each other and unlike charges together? What holds protons, neutrons, and electrons together, and what causes

Hollow metal sphere

Metal comb

Roller

Belt

Field lines

Roller

Van de Graaff Generator

Positive electrostatic charges build up on the outside of the belt as it travels around the rollers, while the inside of the belt is negatively charged. The belt mechanism is inside an insulated post that does not conduct electric charge. A metal comb at the top of the belt transfers the positive charges to the metal sphere.

electrons to jump from your hair to a balloon? The answer is electric force. Electric force, like other forces, can cause a change in the speed, shape, or position of something. Electric force causes the movement of charged particles, such as electrons.

In physics, forces such as electric force can be expressed mathematically. The equation that describes electric force is called Coulomb's Law.

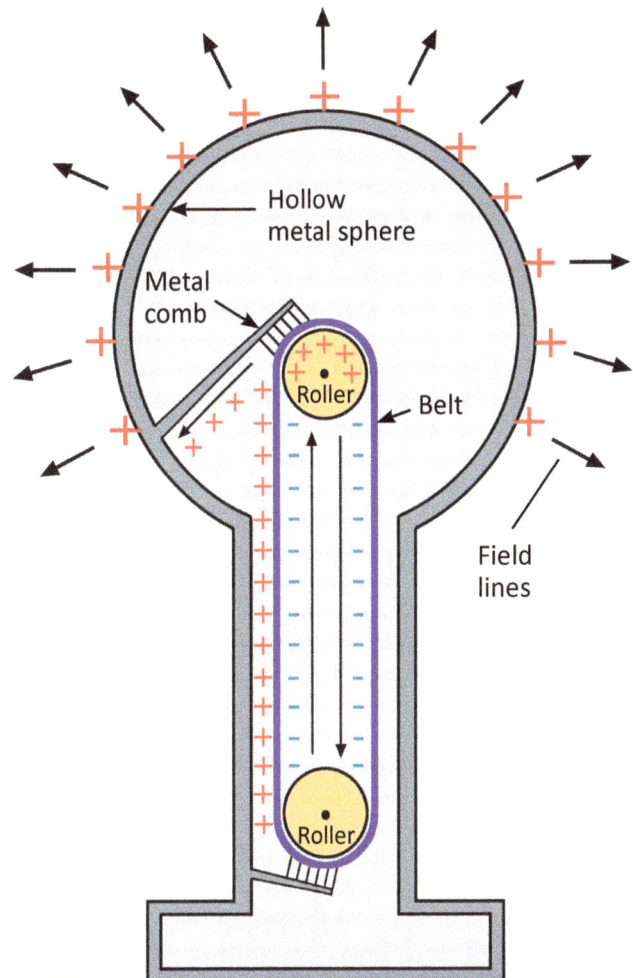

Coulomb's Law:

$$F = k \left[\frac{(q_1\, q_2)}{d^2} \right]$$

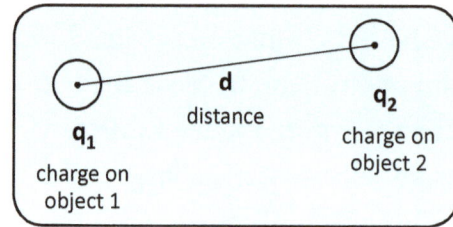

Where F = the force between two objects

q$_1$ and q$_2$ = the quantity of charge on particle 1 and particle 2 respectively

d = the distance between particle 1 and particle 2

k = a proportionality constant that converts the units on the right-hand side of the equation to unit force

$$k = \frac{9 \times 10^9 \text{ Nm}^2}{\text{C}^2}$$

Where N = newton—the unit of measure for force

m = meters

C = coulomb—the unit of measure for electric charge; 1C = the amount of charge of 6.25 billion billion electrons

You can see that "k" is a huge number, which means that the force between particles at very small distances is very large. This is the reason atoms don't fly apart—the forces that are holding atoms together are super strong!

Coulomb's Law Example

How strong is the force between two point charges, each with +1.00 coulomb and 1 meter apart?

Using Coulomb's law:

$$F = k \left[\frac{(q_1\, q_2)}{d^2} \right]$$

q$_1$ = 1.0 C

q$_2$ = 1.0 C

d = 1.0 m

$$k = 9 \times 10^9 \frac{\text{Nm}^2}{\text{C}^2}$$

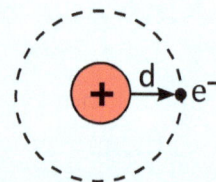

All units but N cancel:

$$F = \frac{(9 \times 10^9\, \frac{\text{N}\cancel{m^2}}{\cancel{C^2}})\,(1.0\,\cancel{C})\,(1.0\,\cancel{C})}{1.0\,\cancel{m}}$$

$$F = \frac{(9 \times 10^9 \text{ N})\,(1.0)\,(1.0)}{1.0}$$

$$F = 9 \times 10^9 \text{ N}$$

d = distance between the electron and the proton

10.5 Electric Fields

When you push a toy car, you apply a force that moves the car forward. The toy car doesn't start moving until you push on it with your hand. In this case the force is transferred from your hand to the toy car through contact. But if you pass a charged balloon over your head, your hair will move towards it without the balloon making contact. Why does this happen?

Certain forces, such as electric forces, act between objects that are not in contact with each other. There is an area around a charged particle where other charged particles can be influenced. This area can be thought of as a kind of force field. For electric charge it is called an electric field. There are also magnetic fields around charged particles if they are moving. We will learn about magnetic fields in the next chapter.

Electric fields have a direction and a strength, or magnitude. Recall from Chapter 7 the terms scalar and vector. Recall that a scalar has a size but no direction and a vector has both size and direction. (See the following page for descriptions of scalar and vector.)

The strength of an electric field is the strength of the force caused by the field per unit charge, which is the charge on one unit—one electron, one molecule, one ion, etc. The strength of an electric field can be written as:

$$E = F/q$$

Where E = electric field

F = the force at some point in space

q = the charge

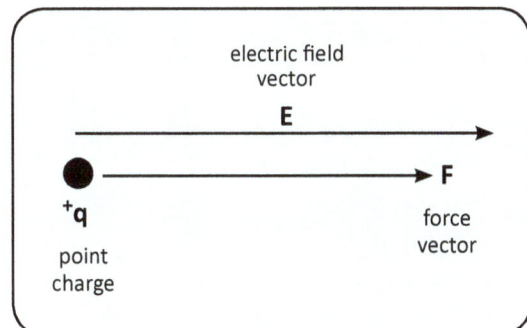

Scalars and Vectors

Mathematically, speed is defined to be a scalar quantity. In math, the word scalar simply means an amount or magnitude. For example, if a car is traveling at 40 km per hour, this describes the *amount* of speed at which the car is traveling from one point to another. A scalar is simply a number that represents a value like speed, temperature, weight, or height. In an equation, a scalar is often written as a lower case letter. For example: s, d, t

Mathematically, velocity is defined to be a vector quantity. In math, a vector has both magnitude (amount) and direction (up, down, right, left, north, south, etc.). A vector is often written as a lower case letter in a **bold** font, such as "**v**" for velocity.

We can think of a vector as an arrow. The length of the arrow represents the magnitude, and the orientation of the arrow represents its direction.

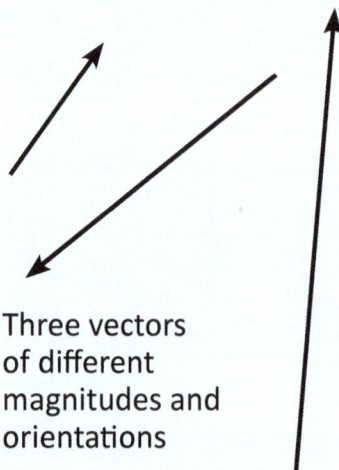

Three vectors
of different
magnitudes and
orientations

Speed (a scalar) is the magnitude (numerical value) for velocity (a vector). The length of the arrow represents the numerical value of the magnitude.

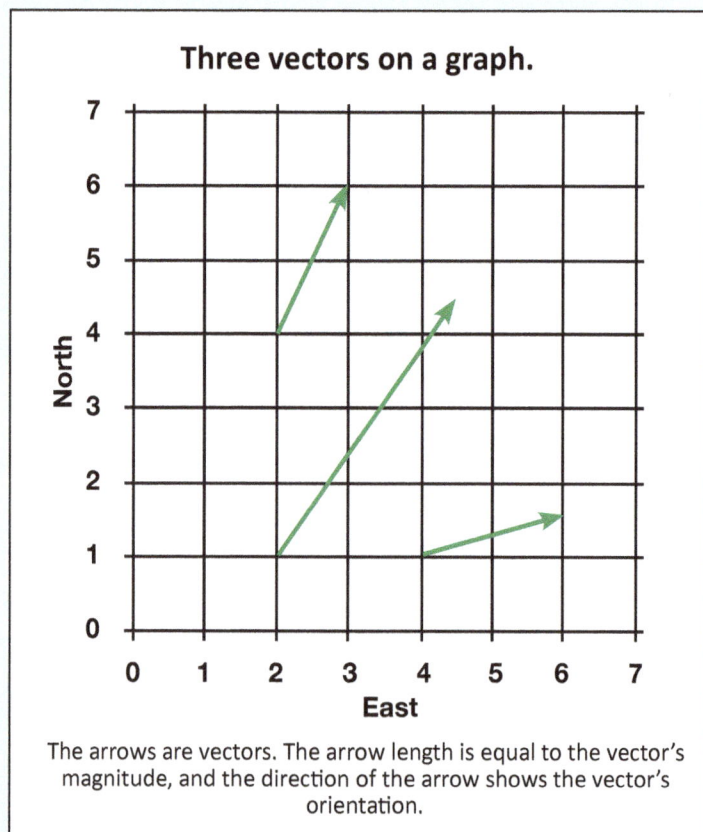

Three vectors on a graph.

The arrows are vectors. The arrow length is equal to the vector's magnitude, and the direction of the arrow shows the vector's orientation.

An electric field emanates radially *away* from a positively charged particle and *inwards* towards a negatively charged particle. If the charges are isolated, the lines extend to infinity. A point charge is the charge at a single point as opposed to a line, a plate, or the surface of a sphere.

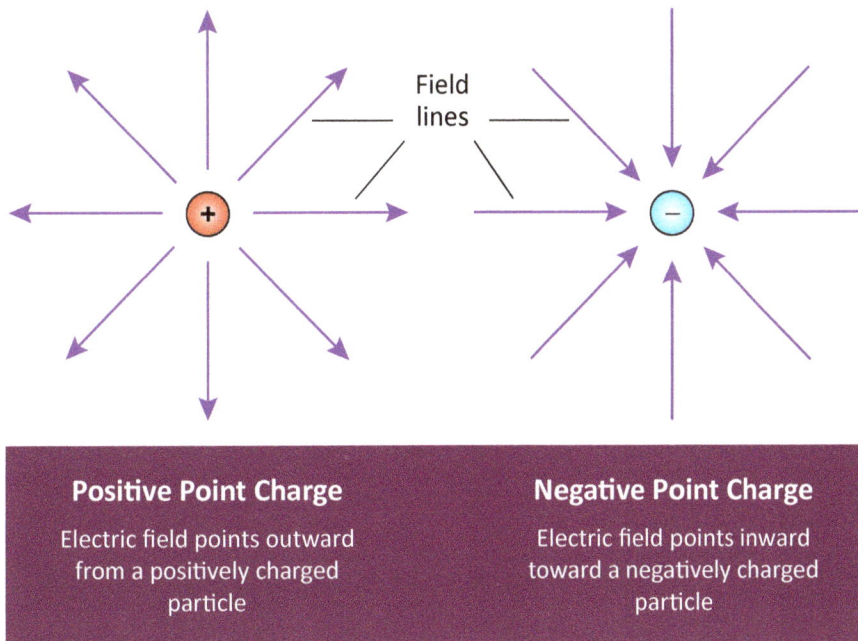

Field lines

Positive Point Charge

Electric field points outward from a positively charged particle

Negative Point Charge

Electric field points inward toward a negatively charged particle

However, if another charge is present, the two charges will superimpose on each other, bending the direction of the electric field lines. For two point charges of opposite charge, the field will look like this:

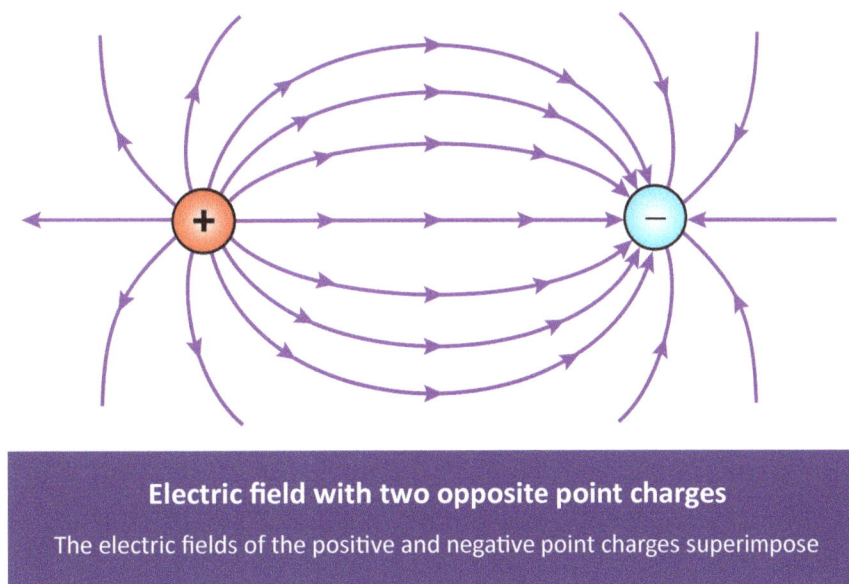

Electric field with two opposite point charges

The electric fields of the positive and negative point charges superimpose

For two oppositely charged parallel plates, the electric field lines will look like this:

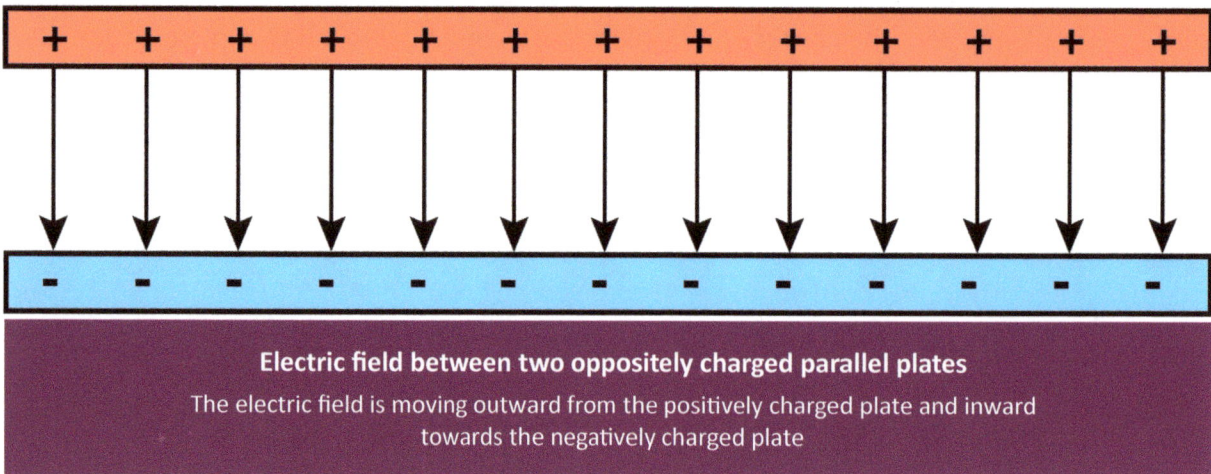

Electric field between two oppositely charged parallel plates
The electric field is moving outward from the positively charged plate and inward
towards the negatively charged plate

10.6 Electric Potential Energy

Recall that an object that is elevated above the ground has the potential to do work. We say that it has gravitational potential energy. For example, if you pick up a book from the floor and place it on a table, the book now has gravitational potential energy. Work is required to lift the book to the table. Recall that work is force times distance, so you did work by applying a force with your hands, lifting the book and then moving the book a given distance from the floor to the table.

A similar argument can be made for electric charge. An electric charge can have electric potential energy, or the potential to do work, depending on its location in an electric field. For example, what happens if you have a single positive charge inside the positively charged electric field surrounding the sphere of the Van de Graaff generator? Because like charges repel each other, if you want to move the positive charge closer to the surface of the sphere, you will have to use a force to push against the electric field which is radiating outward from the sphere. The stronger the electric field, the stronger the force needed to move the single charge toward the sphere and the more work will be needed.

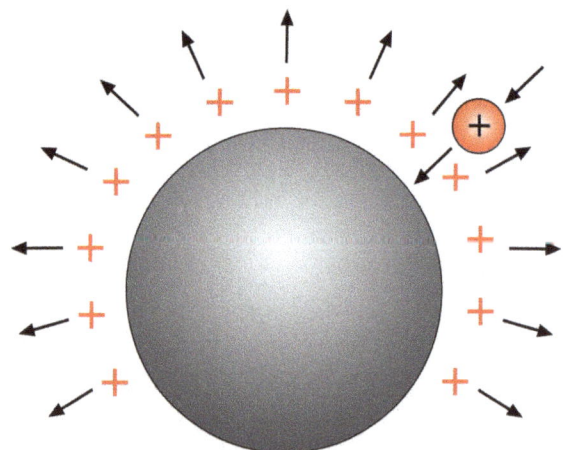

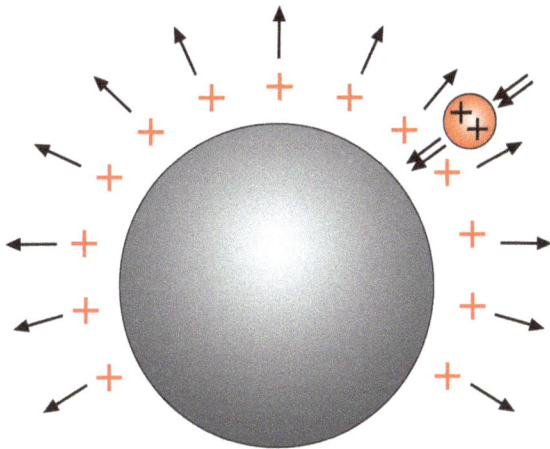

The amount of work is also dependent on the magnitude, or size, of the charge. A particle with twice as much charge would require twice as much work to move it against the electric field and towards the sphere. Recall that work is force x distance. If the force stays the same, then the work is increased or decreased as the distance changes. This is what happens when you move a book higher and higher from the ground—the gravitational potential energy increases. However, for movement through a given distance, the work is greater for a greater force and smaller for a smaller force. Coulomb's Law shows us that if the charges increase, the force increases.

Therefore, a doubly charged particle moving through an electric field gains twice the electric potential energy as a singly charged particle. If the particle has five times the charge, it gains 5 times the electric potential energy, and if it has 100 times the charge, it gains 100 times the electric potential energy. It's like trying to move a bigger and bigger boulder!

10.7 Electrostatic Induction

When you bring a charged object near a conductor, the electrons on the surface will move around until they settle into a new arrangement. The electrons move because they feel forces from the charges on the object.

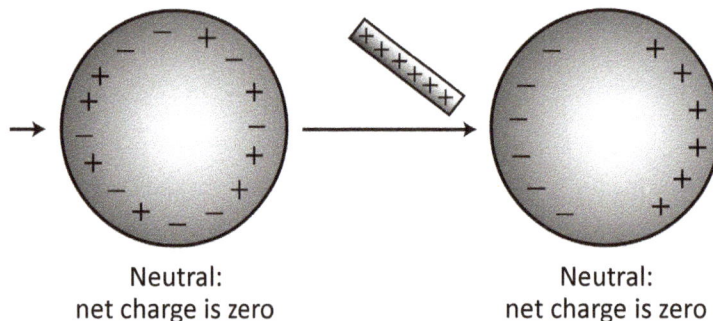

Neutral:
net charge is zero

Neutral:
net charge is zero

If both sets of charges are positive, the charges will move as far away from each other as possible. Charges always spread evenly on a spherical conductor, unless they are pushed by other charges.

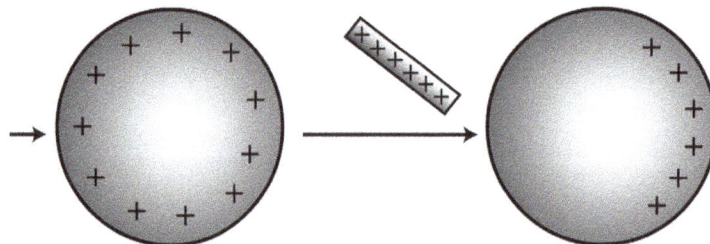

If one set of charges is positive and the other negative, the charges will move so they are closer to each other, but again, the negative charges will repel and won't bunch up.

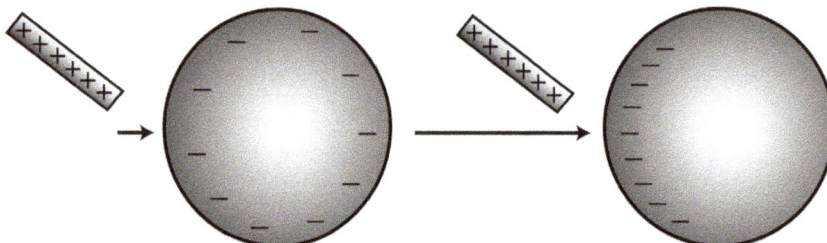

When charges move closer to or farther from a charged object, the effect is called electrostatic induction. Electrostatic induction occurs even though the charged objects don't touch each other. If the objects touch, a path is created for the charges to move back and forth, and this can allow them to neutralize each other, leaving the objects with no net charge.

An electroscope is a simple instrument that uses electrostatic induction to detect electrostatic charge. An electroscope measures static electricity using the fact that like charges repel. Two charged objects with the same charges always move as far away from each other as they can.

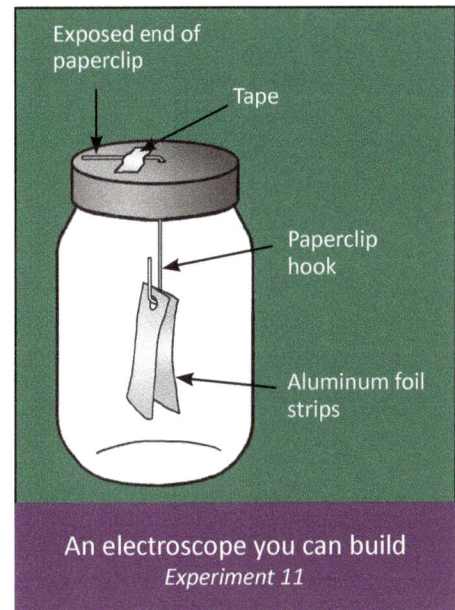

An electroscope you can build
Experiment 11

In *Experiment 10* of the *Laboratory Notebook* you will learn how to make your own electroscope (see illustration). You can build up a charge on a rubber or plastic rod by rubbing it with a silk cloth. When you touch the paperclip on the top of the electroscope with the charged plastic rod, the electrons move as far apart as possible, spreading from the rod along the paperclip and then along the aluminum strips. As a result, both of the foil strips will have negative charge and will repel and move away from each other.

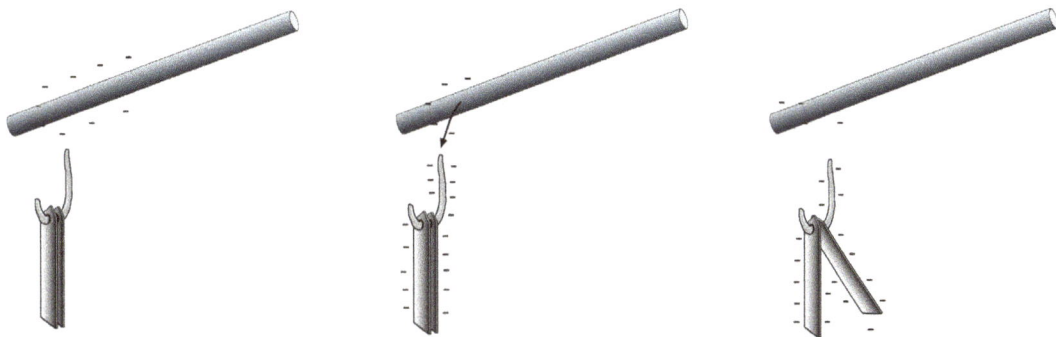

Charged rod before it is touched to the paperclip in the electroscope. The aluminum foil strips are not charged.

The charged rod is touched to the paperclip and the "like" negative charges begin to repel each other, moving from the rod to the foil strips.

The "like" charges on the aluminum foil strips repel, causing the strips to move apart.

The more charge, the more strongly the strips repel; therefore, you can tell how strongly charged the rod was before you touched it to the paperclip. After awhile the charge will leak away, and the foil strips will come back together.

10.8 Summary

● Electrostatics is the study of electric charges at rest. Electric charges at rest are called static electric charges or, commonly, static electricity.

● A positive ion is an atom that has a net positive charge because it is has lost one or more electrons, and a negative ion is an atom that has a net negative charge because it has gained one or more electrons.

● As electrons from an object are added or removed, a slight imbalance in charge occurs, and this imbalance will cause the object to have an electrostatic charge.

● Electrical force causes the movements of charged particles, such as electrons.

● An electric field is the area force field per unit charge around a charged particle that can influence other charged particles.

● Electrostatic induction occurs when a charged object causes the rearrangement of charges on another object.

10.9 Some Things to Think About

● Which of these is not the result of electrostatic charges?

Hair standing on end.

A balloon sticking to a wall.

Clothes sticking together when taken out of the dryer.

A flashlight giving out a beam of light.

Getting a shock from your cat.

● What do you think holds atoms and molecules together?

● Why do you think some objects are charged and others are not?

- Why do you think the air needs to be dry for you to be able to generate an electrostatic charge using a balloon and your hair?

- Explain what keeps like charges away from each other and unlike charges together.

- Describe an electric field.

- What happens to an electric field when there are two oppositely charged particles near each other?
  When there are two oppositely charged plates?

- If you doubled the charge on a particle, how would this change the amount of work needed to move it?

- What happens to force as charges increase?

- How does electrostatic induction work?

Chapter 11 | Electrodynamics

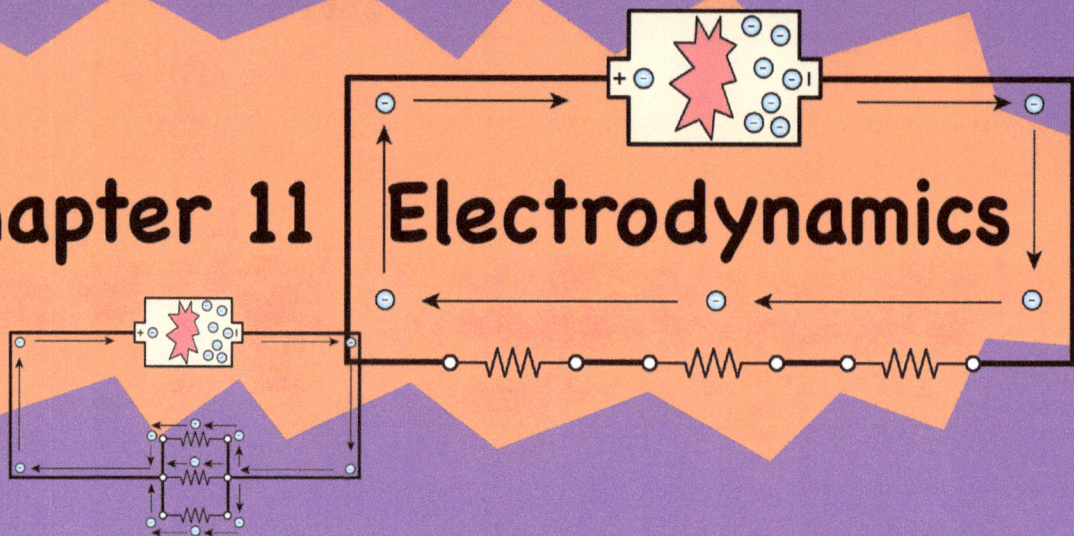

11.1 Introduction

In the last chapter we learned about electrostatics, which is the study of electric charges at rest. We saw that like charges will repel each other and unlike charges will attract each other. We also saw that an electric force can move electrons from one object to another.

Electric charges can also *flow*. In fact, electric charges flow in wires much like water flows in a garden hose. When you plug in your radio, it's like hooking a garden hose to a faucet. When you turn on your radio and you hear music, it's like turning on the faucet and letting the water go through the garden hose. The electric charges flow from the outlet through the wire to your radio, just like water flows through a garden hose. Flowing electric charge is called electric current.

Before we look at electric current, let's take a closer look at how water flows out of a garden hose. What happens in a garden hose when you turn the water off? Does water keep flowing through the hose? No. A few leftover drops may fall out, but water does not continue to flow through the hose. This happens because there is nothing pushing the water out of the hose. To make the water flow through and out of the hose, something needs to be pushing on the water. That "something" is pressure. Pressure forces the water out of the hose, and without pressure the water will not flow. When we take a closer look at how water pressure works, we can better compare water pressure to electrical pressure.

Water towers— the water in the taller tower has more gravitational potential energy than the water in the shorter tower

Recall that pressure is force per unit area. One way to create water pressure is to place a water tank at the top of a tall tower. Recall that any object raised above ground level has gravitational potential energy. Due to the force of gravity, the water at the top of the tank pushes against the water at the bottom of the tank creating pressure. You can see that the taller the tank, the greater the gravitational potential energy. This creates more pressure which in turn makes the flow of water stronger.

When a pipe is attached to the bottom of the tank, the pressure will push water through the pipe. If this pipe is attached to the pipes in your house, the pressure created in the water tank will push water through the pipes in your house and through your garden hose whenever a faucet is open. However, if the faucet is turned off, the water cannot flow. With the faucet off, there is only the *potential* for water to flow.

When the faucet is open, water with higher pressure (high force per unit area) is pushing against water with lower pressure (low force per unit area), forcing the water out of the opening in the faucet. Notice that without water pressure from the water tower or a pump and without water pressure in the plumbing of your house, there would be no *potential* for water to flow. The water pressure creates the *potential* for water to flow through the garden hose.

Also notice that in order for water to flow through it, the hose needs to be open at one end. Another way to say this is that one end of the hose must have low or zero pressure (the open end of the hose has zero pressure). When the faucet is open and the far end of the

hose is also open, the high pressure from the water in the tank will push the water through the hose, creating a continuous flow of water. So a garden hose needs **both** a high pressure source to push the water through the low pressure end of the hose and a low pressure space for the water to be pushed into.

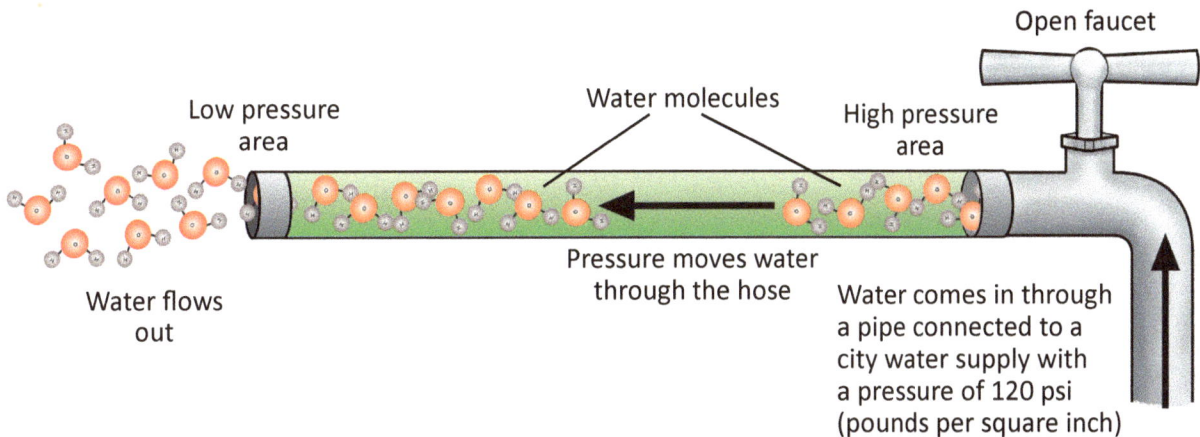

Low pressure area

Water molecules

High pressure area

Open faucet

Water flows out

Pressure moves water through the hose

Water comes in through a pipe connected to a city water supply with a pressure of 120 psi (pounds per square inch)

Water under high pressure moves through a pipe and a hose

The same is true for electric charges. In order for electric charges to flow through a wire, there must be some sort of pressure. This pressure is called electric potential, or voltage. Let's take a closer look at voltage (electric pressure). Electric potential is "pressure" for electric charge.

You are probably familiar with the word voltage, but you may not have heard it being called electric potential. Let's compare how electric charges flow through a wire to how water flows through a garden hose. As we just saw, water moves through the hose when water molecules that are under higher pressure push against molecules that are under lower pressure. This causes the water molecules to move forward through the hose. The same is true for electrical potential. In a wire, electrons that are under higher electrical pressure hop from atom to atom displacing other electrons that are under lower electrical potential. The displaced electrons then hop to new atoms. In this way electrons flow through a wire, hopping from atom to atom.

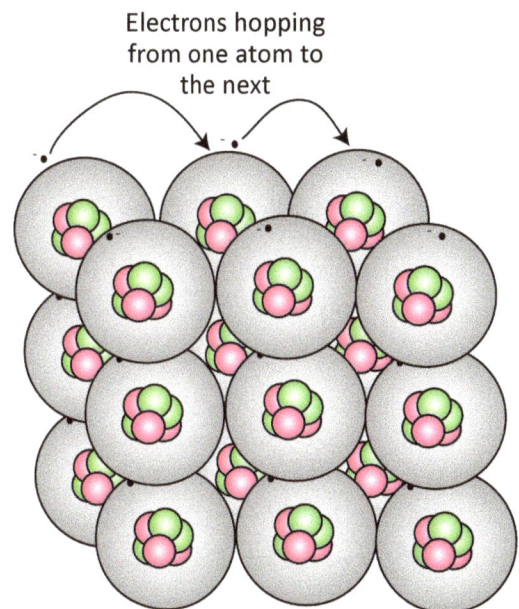

Electrons hopping from one atom to the next

Using the garden hose analogy, we can begin to understand how the terms electrical pressure, electric potential energy, and voltage are the same. For example, a battery has the *potential* to create electrical pressure through chemical reactions. Some batteries have the *potential* to create more electrical pressure than others. The potential to create electrical pressure is called electric potential. The unit of measure for electric potential is the volt. The voltage, or number of volts, of a battery is a description of the amount of electrical pressure it has the *potential* to produce. In other words, a 1.5 volt battery can produce 1.5 volts of electric pressure and a 12.5 volt battery can produce 12.5 volts of electrical pressure.

The "high pressure side" of a battery is called the anode (-) and the "low pressure side" of the battery is called the cathode (+). An electrochemical process within a battery creates electrons that flow toward the anode, and the electrons build up there. This creates an electric potential or electric pressure. Like water, electrons will flow from a higher pressure area to a lower pressure area.

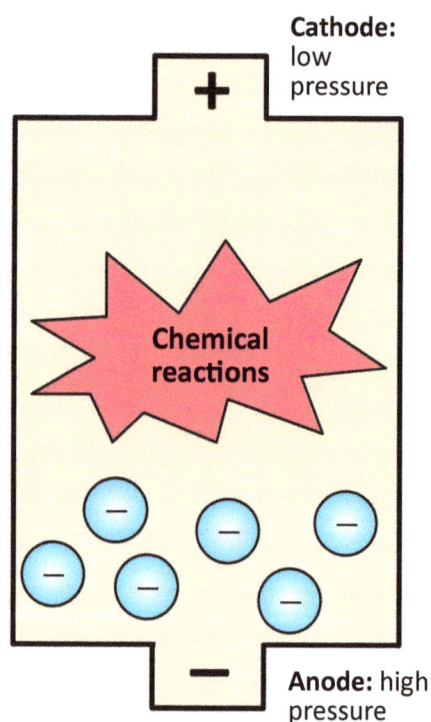

Cathode: low pressure

+

Chemical reactions

−

Anode: high pressure

When a conductive wire is attached to the anode at one end of the battery and the cathode at the other end, the electrons can flow out through the anode, through the wire, and back to the other side of the battery through the cathode. This path is called a circuit. As long as the circuit has a conductive wire connected between the anode and the cathode, the electrons will flow. This is called a closed circuit. If the circuit has a break somewhere, it's called an open circuit. Having a break in a circuit is like turning off the faucet. The electrons cannot flow. We'll learn more about circuits in Section 11.4.

When we measured the voltage of our battery in *Experiment 9* of the *Laboratory Notebook,* we were measuring how much electrical pressure it produced. Batteries with large voltages have large electrical pressures and can move more electrons through wires than batteries with smaller voltages and less electrical pressure. However, unlike a hose that has water moving through it from a faucet, the electrons are already in the metal wire. When you plug your radio into an electric outlet in your home, you are providing electrical pressure to move the electrons that are already in the wire.

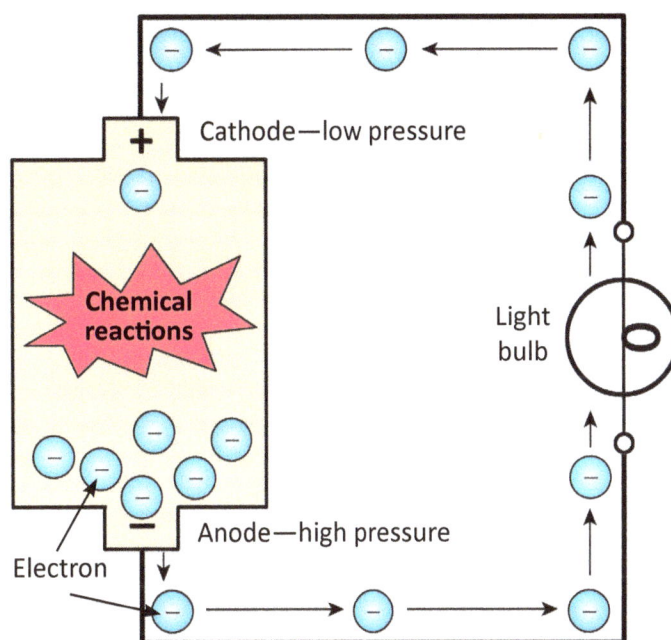

Cathode—low pressure

Chemical reactions

Light bulb

Electron

Anode—high pressure

A closed circuit

Electric potential (pressure) from chemical reactions forces electrons to flow from a "high pressure" anode through a wire and into the "low pressure" cathode, creating a continuous flow of electrons.

11.2 Conductors and Insulators

Only the electrons can move through a conductor. Protons cannot move. Recall that electrons orbit the nucleus of the atom and protons stay inside the nucleus. Some of the electrons that orbit an atom are free to move, but protons are "locked" and are not free to move. Protons stay firmly in place in the nucleus. However, electrons *can* hop from atom to atom, going in and out of the orbits of the atoms in certain materials. Materials that allow electrons to move are called conductors. Materials that do not allow electrons to move are called insulators.

Conductors

Materials that allow electrons to flow freely from atom to atom are called conductors. Conductors include all metals, some non-metals such as graphite, saltwater solutions and other electrolytes, and a whole host of superconductor and semiconductor materials.

Not all conductors have the same level of conductivity. Silver is the best conductor and allows the most free (easiest) passage of electrons, but if you cut open an electrical cord, you will usually see a group of copper wires wrapped tightly together inside the plastic covering. Copper is used rather than silver for electrical components and wiring in houses and electrical appliances because copper is cheaper to produce and is more malleable than silver. Although copper is slightly less conductive than silver, it still has very high conductivity. Gold has less conductivity than either copper or silver but is used in some small-scale projects because gold is nonreactive and won't corrode like silver and copper. Corrosion occurs when oxygen in the air reacts with the metal, which can reduce the conductivity. Since it is nonreactive, gold requires no protection from the atmosphere.

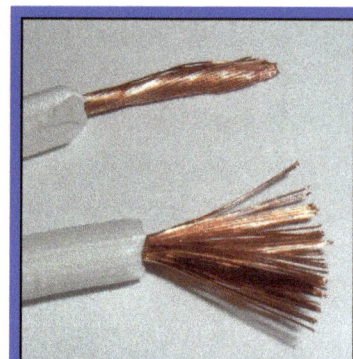

Copper wire is a good conductor

Photo of copper wire in a lamp cord courtesy of Scott Ehardt

Increased conductivity →

Silver
Copper
Gold
Aluminum
Tin

The size and shape of a conductor will also affect its ability to conduct. For example, if we have two wires made of the same material and one wire is thick and the other wire is thin, the thicker wire will conduct more easily than the thinner wire for the same length. Going back to the example of a garden hose, if you have two hoses of the same length, a fat hose will allow more water to be pushed through than a thin hose because the fat hose has more space in it.

The same is true for a longer wire and a shorter wire of the same thickness. The shorter wire will allow easier passage of electrons than the longer wire because the electrons have less distance to travel.

Insulators

Insulators are materials that don't allow movement of electrons from atom to atom. Why do some materials allow electrons to flow and others don't? Why are some materials conductors and others insulators? As it turns out, some materials are resistant to allowing electrons to flow from atom to atom. The word resistant comes from the Latin word *resistere*, which means "to stand [still]." So, to resist something means to stand against, or refuse, to do something.

I AM RESISTANT TO BRUSSELS SPROUTS!

Are you resistant to eating brussels sprouts or spinach casserole? If so, there may be many reasons why you don't want to eat these foods. Maybe you don't think they taste good, but another reason might be that you are just *too full!* The same is true for electrical resistance. Insulators, such

Insulators

Glass
Rubber
Oil
Fiberglass
Ceramic
Plastic

Increased electrical resistance

Insulators

1. Insulator on high voltage power lines supports the wires without conducting electrical current;
2. Three cables made of twisted copper wire, each covered with a different color plastic insulation;
3. Cable with two conducting wires surrounded by insulating material

Photo credits, Wikimedia Commons: 1. Adrian Pingstone (Public Domain); 2. Chatama CC BY SA 3.0; 3. Alistair1978- CC BY SA 2.5

as foam or plastic, won't allow electrons to move from atom to atom because they don't have extra space in their atoms to accommodate new electrons. Their atoms are "too full."

Just like conductors have different abilities to conduct electrons, insulators have different levels of electrical resistance because some have atoms that are "full" and others have atoms that are only partially "full." This makes some insulators more effective at blocking electron flow than other insulators.

11.3 Semiconductors and Superconductors

The classification of a material as a conductor or an insulator depends on how tightly the atoms hold on to their electrons. There are some materials that are neither good conductors nor good insulators but fall in between the two. For example, germanium and silicon are neither good insulators nor good conductors. However, they can be pushed towards having either more insulating or more conducting properties by adding impurities or even exposing them to light. Another example of a material that can change properties is selenium which is normally a good insulator but becomes a conductor if exposed to light!

Materials that can behave as insulators some of the time and conductors some of the time are called semiconductors. Semiconductors are very useful for building circuits because they can be used to control the flow of electric current. A transistor is a component in electric circuits that is made of layers of semiconducting material sandwiched together. In addition to controlling the flow of electric current, transistors can also detect and amplify radio signals and act as digital switches, turning the electric flow on and off, just like a faucet attached to a garden hose.

Some materials change their electrical properties when they undergo different amounts of heating or cooling. For example, glass is a very good insulator at room temperature, but heat some types of glass to a very high temperature and they become a conductor! Most metals are better conductors when cold than when heated.

When some materials are supercooled, or cooled to very low temperatures of about -140°C to -190°C, they become superconductors. Superconductors conduct electrons perfectly with no electrical resistance. If a closed loop of superconducting wire gains a current, the current never stops. The electrons can continue flowing around the loop forever. Most superconductors are supercooled metals but new high temperature superconductors are being developed. The temperatures are still low (around -70°C) but much higher than temperatures for previous superconductor materials. Scientists are working to find materials that will superconduct at room temperature.

Imagine riding on a train that uses superconductors to make it float ever so slightly off the rails! This train could go super fast and be super quiet and super smooth. In fact, in 1962 research began in Japan to test the concept of maglev (magnetic levitation) trains that use superconductors to levitate them above the tracks. By 2015 after many tests of different machines, a prototype train was developed that was able to go 603 km/h (374 mph)!

A superconductor levitating a magnet

Small electrical currents flow from the superconductor causing the magnet to be repelled.

Courtesy of Mariusz.stepien, Wikemedia Commons, CC BY SA 3.0

Japan is currently planning a maglev route from Tokyo to Nagoya which they hope to have operational by 2027. People would be able to take a speedy 40 minute trip by maglev train that would take almost 5 hours by car. Meanwhile, the first commercially used maglev train was put into service in Shanghai, China in 2004. The Shanghai Maglev Train can travel at 431 km/h (268 mph) and goes a distance of 30.5 km (19 mi) in about 8 minutes. If we can discover how to make superconductors at room temperature, it would help us develop efficient maglev trains all over the world.

A Japanese maglev train prototype

Courtesy of Daylight9899, Wikemedia Commons License CC by SA 3.0

11.4 More About Circuits

Recall that a circuit is any path that connects two points and through which an electrical current can flow. There are two type of basic circuits— series circuits and parallel circuits.

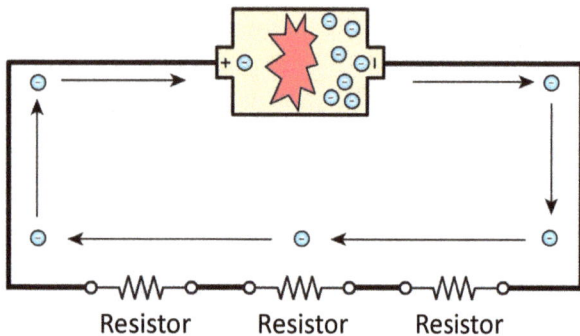

A series circuit
All the components are connected along a single path.

In a series circuit all of the components including the power source, conductors, and resistors are connected along a single path, one after the other. In a series circuit electrons have only one path they can take. If a series of resistors is attached to the wire between the anode and the cathode, the electrons will flow along a path from the anode, through the resistors, and then to the cathode.

In a parallel circuit the components are in parallel, with their heads connected together and their tails connected to each other. Some components, such as diodes, batteries, and transistors, have "polarity" with "heads" and "tails." Other components, such as resistors and capacitors, usually do not have "polarity." This allows electrons to flow through different paths at the same time.

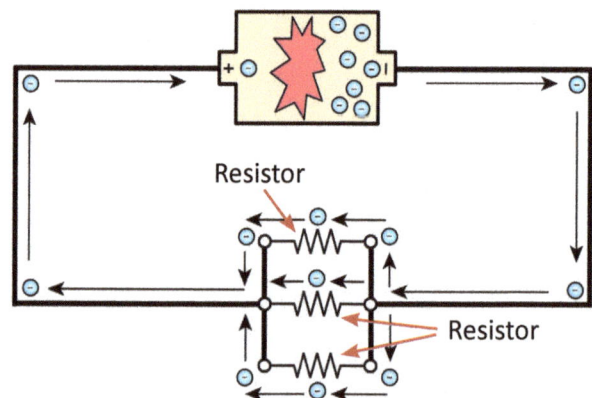

A parallel circuit
All the components are connected in parallel. Electrons flow through different paths at the same time.

11.5 Ohm's Law

In 1825 a German mathematician and physicist named Georg Ohm (1789-1854 CE) discovered the relationship between voltage, current, and resistance. In 1827 Ohm published his work in a famous book entitled *Die galvanische Kette, mathematisch bearbeitet (The Galvanic Circuit Investigated Mathematically)*. In this book Ohm set forth a complete theory of electric current. Although he was able to show how current, voltage, and resistance are related in a mathematically elegant way, he was not able to convince some of the older German physicists that his mathematical approach was correct. Although his work was initially met with little enthusiasm, in 1841 Ohm was awarded the Copley Medal, which is an award made by the Royal Society of London for outstanding achievement in scientific research. Ohm became a foreign member of the Royal Society in 1842 and in 1845 was also made a full member of the Bavarian Academy of Sciences and Humanities which recognizes scholars whose research has made a major contribution to knowledge within their field of study.

GEORG OHM
Circa 1789-1854

Ohm figured out the mathematical relationship between voltage, current, and resistance, and this relationship is know as Ohm's law.

Ohm's law:

$$I = \frac{V}{R}$$

Where I = current
V = voltage
R = resistance

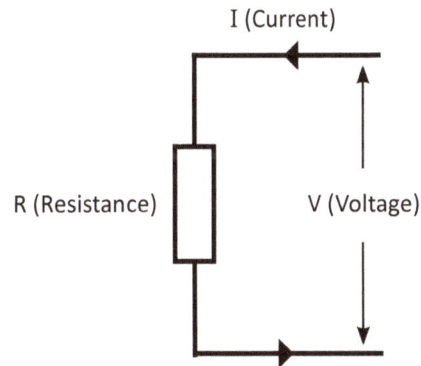

I (Current)

R (Resistance)   V (Voltage)

We can write Ohm's law in unit form (in the form of units of measure) as:

$$\text{Amperes (A)} = \frac{\text{Volts (V)}}{\text{Ohms } (\Omega)}$$

Where ampere is a unit of current (given the symbol A)
volt is a unit of voltage [electric potential] (given the symbol V)
ohm is a unit of resistance (given the symbol Ω—the Greek letter called omega)

Ohm's law tells us that if we have a difference in electric potential "across a circuit" (between any two points on the circuit) of 1 volt and a resistance of 1 ohm, we will get a current of 1 ampere.

$$I = \frac{V}{R} \Rightarrow I = \frac{1 \text{ volt}}{1 \text{ ohm}} = 1 \text{ ampere}$$

If we have 10 volts, we'll get 10 amperes, and if we have 100 volts, we'll get 100 amperes.

$$I = \frac{V}{R} \Rightarrow I = \frac{10 \text{ volts}}{1 \text{ ohm}} = 10 \text{ amperes}$$

$$I = \frac{V}{R} \Rightarrow I = \frac{100 \text{ volts}}{1 \text{ ohm}} = 100 \text{ amperes}$$

Strange Names!

Where do we get these names for units of measure?

Each unit is named after the scientist who developed a different aspect of electrical theory.

These famous scientists are: Andre M. Ampere (French), Alessandro Volta (Italian), Georg Ohm (German).

A Toasty Example

Sample problem

A toaster carries a current of 4.0 A (amperes) and is connected to a 120 V (volt) source. What is the resistance of the circuit?

Solution

$$I = \frac{V}{R} \Rightarrow \text{solve for R} \Rightarrow R = \frac{V}{I}$$

$$R = \frac{120V}{4A} = 30 \, \Omega$$

Why did you learn at a young age not to put objects like your finger into an electrical outlet? Why did your parents put covers on the outlets when you were young? Covering outlets helps prevent harmful accidents. If too much electric current passes through a human body, it can be quite painful and can even cause death. It only takes a small amount of electric current to cause great damage to the human body.

The resistance in your body can be as low as 100 ohms if you have soaked in salt water or as high as 500,000 ohms if you live in a dry climate and your skin is very dry.

Shocking Electric Currents

Current in amperes	Effect on the Human Body
0.001 A	Can be felt
0.005 A	Is painful
0.015 A	Causes loss of muscle control
0.070 A	Can be fatal

Adapted from *Conceptual Physics* by Paul Hewitt

A Very Shocking Problem

Sample problem

Assume your body has a resistance of 200,000 ohms and you touch both terminals of a 9-volt battery. How much current will pass through your body?

$$I = \frac{V}{R} = I = \frac{9 \text{ volts}}{200,000 \text{ ohms}} = 0.000045 \text{ amperes}$$

Can you feel the current?   Yes   (No)

Now imagine you have soaked in a saltwater bath and the resistance of your skin goes down to 1000 ohms. How much current will pass through your body?

$$I = \frac{V}{R} = I = \frac{9 \text{ volts}}{1000 \text{ ohms}} = 0.009 \text{ amperes}$$

OUCH!

11.6 Summary

- Electric charges can flow, creating electric current.

- Electric potential pushes electric charges through a wire. Electric potential is called voltage.

- Conductors are materials that allow electrons to flow from atom to atom.

- Insulators are materials that don't allow electrons to flow from atom to atom.

- Resistance is the tendency of a material to not allow electrons to flow from atom to atom.

- Ohm's law is the mathematical relationship between voltage, current, and resistance.

- Units of measure:

  ampere is a unit of current (given the symbol A)

  volt is a unit of voltage (given the symbol V)

  ohm is a unit of resistance (given the symbol Ω)

11.7 Some Things to Think About

- How would you define electric current?

- What makes electric charges flow through a wire?

- If you had an electric current flowing through a circuit, how would you stop the current?

- Explain the difference between conductors and insulators and why they work the way they do.

- If you wanted to control the flow of electric current in a circuit, what components would you add to the circuit? Why would you choose these?

- What is the advantage of superconductors? What problem needs to be overcome for superconductors to be widely used?

- If you were building electric circuits, when do you think you would use series circuits and when would you use parallel circuits?

- What is Ohm's Law?

- Why do you think Ohm's Law was an important discovery?

Chapter 12 Magnetism

12.1 Introduction

Magnets are fascinating to play with. They seem almost magical because they can move objects from a distance. Grab a strong magnet and place it under a table and watch nails wiggle as you move the magnet below.

Magnets have abundant uses and can be found in all shapes and sizes. Tiny magnets are used in computers to read and write data on a hard drive, and large magnets are used in junk yards to lift and move old, decaying cars. Magnets can be used to hold grocery lists on refrigerator doors or snap toy letters into place. Magnets can be used in a variety of ways that help science, medicine, and commerce in the modern world.

Legend has it that over 2000 years ago in the Greek province of Magnesia a shepherd named Magnes was herding his sheep and noticed that when he walked on a certain rock the nails on his sandals and the metal end of his staff were attracted to the rock. He dug up more of these magnetic rocks and found what came to be called lodestones that contain magnetite. The words magnet and magnetite come from the Greek phrase *magnetis lithos* which means "the stone of Magnesia."

Magnetite crystals
(on feldspar)

Courtesy of Rob Lavinsky,
iRocks.com, CC BY SA 3.0

Magnetite is a mineral made of iron oxide (Fe_2O_4). Magnetite varies in color from dark gray to black and forms octahedral crystals like the one in the photograph. It is the only mineral that is a natural magnet. Magnetite forms a permanent magnet—unlike some other materials, once it becomes magnetized it stays magnetized permanently.

Permanent magnets can also be made of iron and of rare earth metals like strontium and samarium. Strong magnets made of samarium and cobalt can produce magnetic forces 50 times stronger than magnetite, and magnets made from neodymium, iron, and boron can produce magnetic forces almost 75 times stronger than magnetite.

A toy made of rare earth magnets (neodymium)

Courtesy of Cskey, CC BY SA 3.0

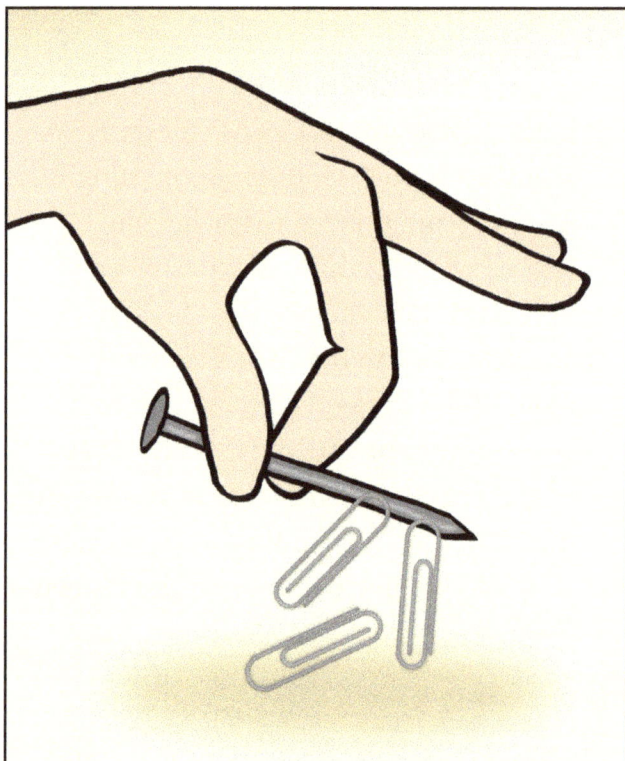

Instead of being permanent, some magnets are only temporarily magnetic. Induced magnetization occurs when a material such as iron becomes temporarily magnetized by a nearby magnet. It is possible to use a permanent magnet to magnetize an iron nail, for example. If an iron nail is brought close enough to a permanent magnet, the iron nail will become magnetized. An iron nail can remain magnetized, or over time it can lose its magnetism depending on how its magnetic domains are aligned. We will learn more about magnetic domains in the next section.

Why are some materials magnetic and others not magnetic? And how can a permanent magnet make another object temporarily magnetic? In the next several sections we'll answer these questions as we take a closer look at magnetic fields, magnetic domains, and electromagnets.

12.2 Magnetic Fields

In previous chapters we looked at electric forces in electrons and how moving electrons create electric fields when they flow through a wire. Another type of movement of electrons creates magnetic forces and magnetic fields. A magnetic field is the area around a magnet that is affected by magnetic force. All magnetic fields are the result of moving electric charges. As we saw in the last chapter, electrons can move by hopping from atom to atom to produce electric currents. In most magnetic materials the magnetic field is caused by electron spin of unpaired electrons. Although there is no evidence that an electron literally spins, visualizing electrons in this way helps us to understand how magnetic forces work.

In 1820 Hans Christian Oersted (1777–1851 CE), a Danish physicist, performed an important experiment that showed that there is a connection between moving electric charges and magnetic fields. Oersted found that when an electric current is passed through a wire, the needle on a nearby compass will move. The needle moves because the electric charges moving through a wire create a magnetic field around the wire. Oersted had discovered electromagnetism. The unit for magnetic field strength, or intensity, is called the oersted in honor of his discovery.

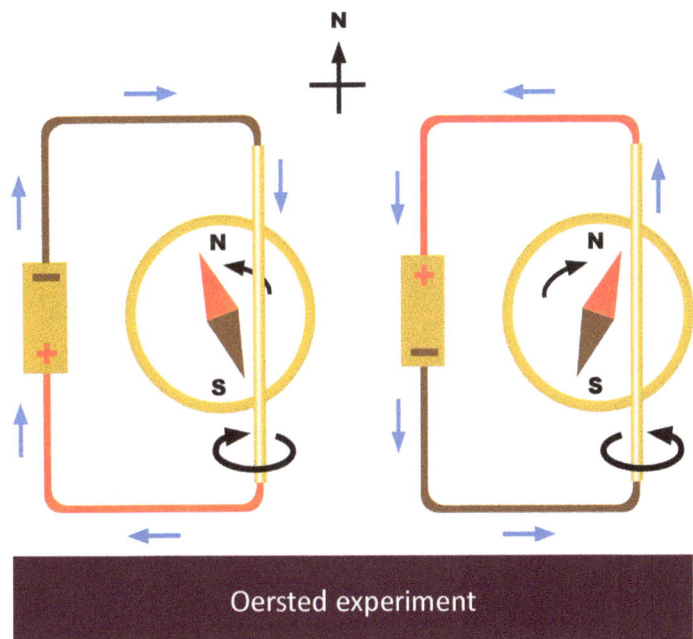

Oersted experiment

If moving electric charges create magnetic fields, why is a mineral like magnetite magnetic? Magnetite and other magnets don't have nearby wires with moving electric charges, yet they are magnetic. How does this work?

It turns out that magnetic fields are created by electrons not only when they move through a wire but also when they spin. In every atom of every material, electrons not only move in an orbit around their nucleus, but they also spin on their axis. When an electron spins, it creates a moving electric charge, and as Oersted discovered, these moving electric charges create magnetic fields. Each atom has a tiny magnetic field. Much like spinning a basketball on your finger, an electron can spin to the right or to the left on its central axis.

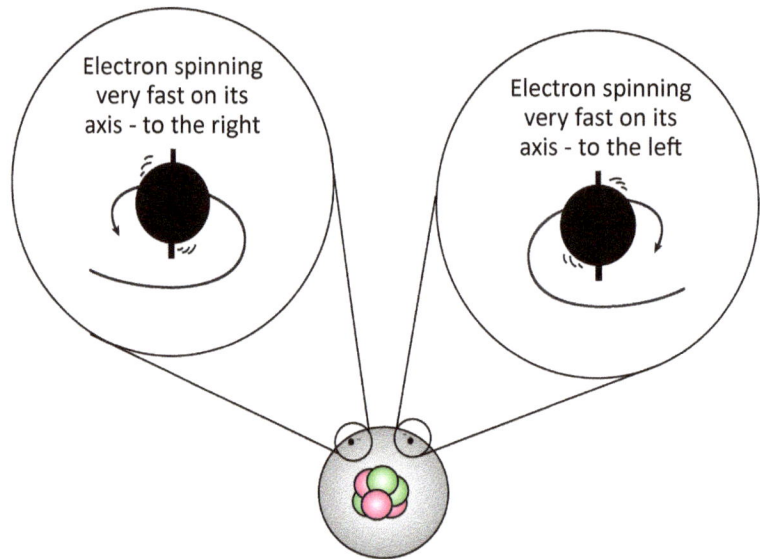

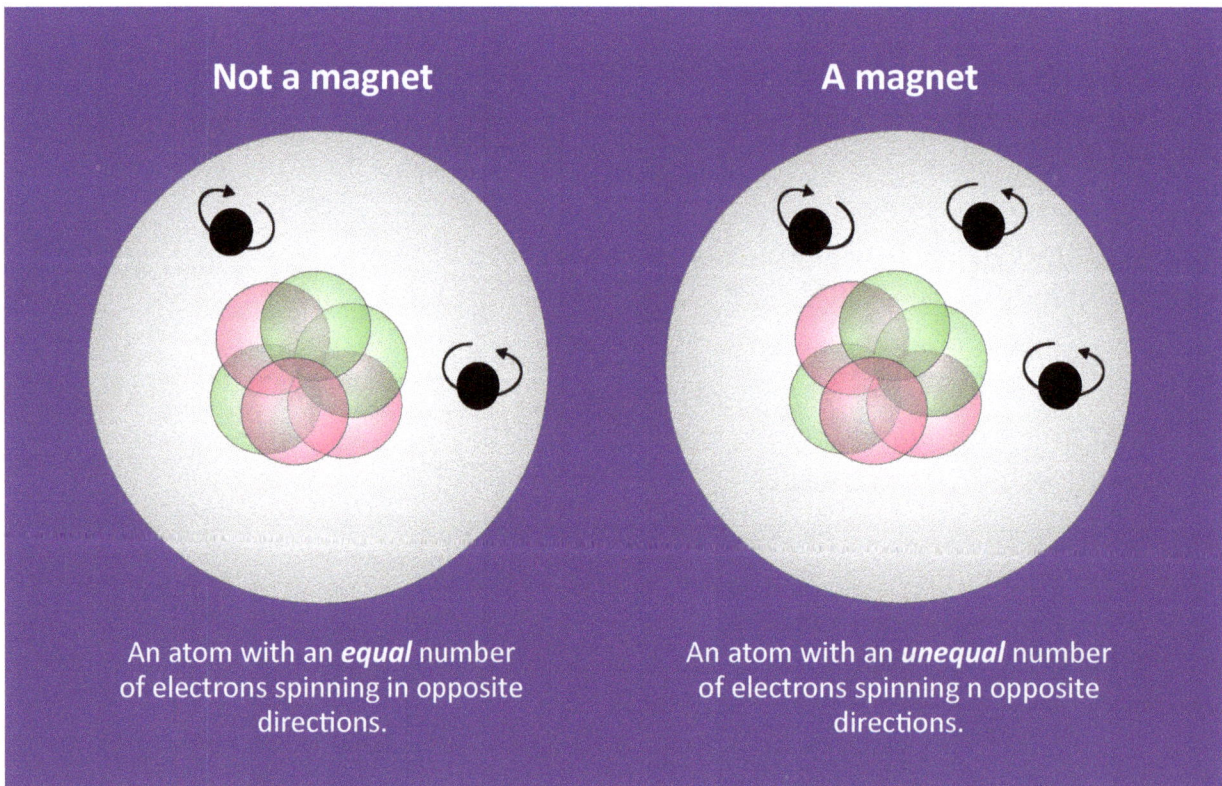

Electron spinning very fast on its axis - to the right

Electron spinning very fast on its axis - to the left

Not a magnet

A magnet

An atom with an *equal* number of electrons spinning in opposite directions.

An atom with an *unequal* number of electrons spinning n opposite directions.

Because the electron spins, it creates a tiny magnetic field. This tiny magnetic field will have a "north" pole and a "south" pole, making it a magnetic dipole. Recall that *dī-* means two—a magnetic dipole has two poles.

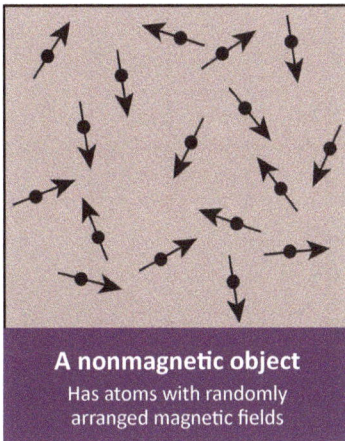

A nonmagnetic object
Has atoms with randomly arranged magnetic fields

If an even number of electrons are spinning in opposite directions, the magnetic fields will cancel and the atom will have no net magnetic field. In nonmagnetic materials the opposite electron spins cancel each other out. However, if an unequal number of electrons are spinning in opposite directions, the extra spinning electron will generate a net magnetic field on the atom.

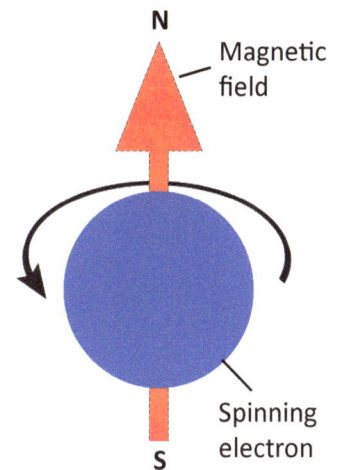

In most materials the atoms are randomly oriented and their magnetic fields cancel each other out. However, materials such as iron, nickel, and cobalt have areas where groups of atoms have magnetic dipole fields that are aligned to form magnetic domains. Magnetic domains are created when individual atoms interact with nearby atoms causing the magnetic fields in a cluster of atoms to line up with each other. Magnetic domains are small, ranging from 10^{-6} to 10^{-4} meters in most materials, but each domain contains billions of aligned atoms.

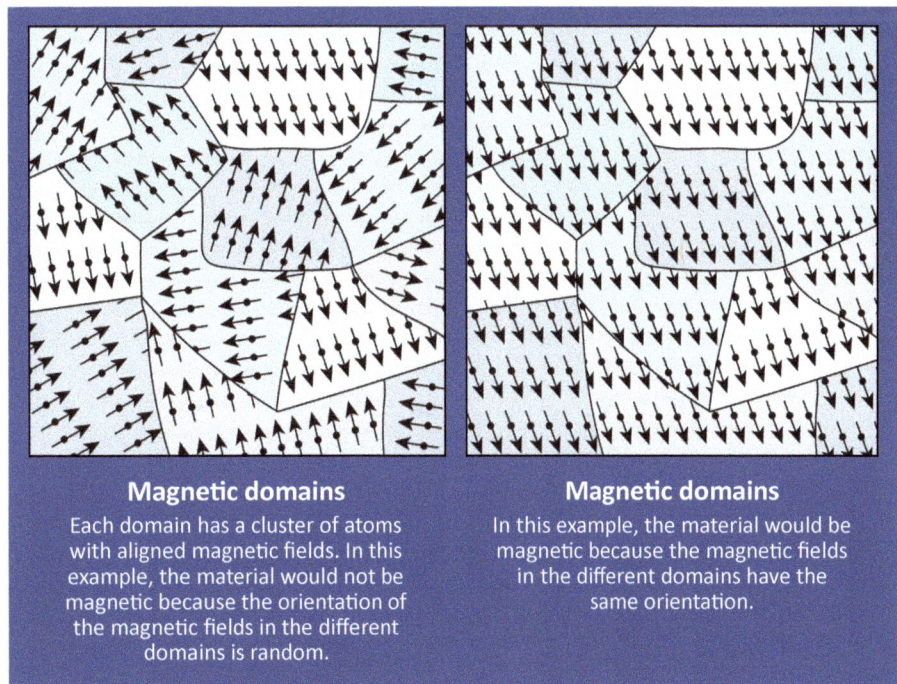

Magnetic domains
Each domain has a cluster of atoms with aligned magnetic fields. In this example, the material would not be magnetic because the orientation of the magnetic fields in the different domains is random.

Magnetic domains
In this example, the material would be magnetic because the magnetic fields in the different domains have the same orientation.

In an ordinary iron nail the magnetic domains are randomly oriented. The nail is not magnetic because the magnetic fields from each domain cancel each other out. However, if a strong magnet is brought close to the iron nail, all of the tiny magnetic domains will begin to align. When enough tiny domains have aligned, the whole iron nail is temporarily magnetized and will be able to attract or pick up objects like paperclips and other nails.

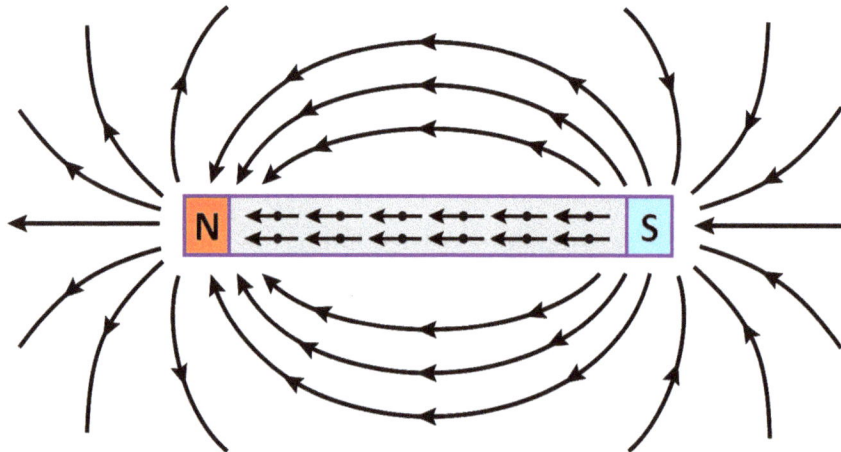

Unlike protons and electrons, magnets don't have a positive or negative charge. Instead, a dipole magnet has a north and a south pole. Looking at the illustration of the aligned magnetic domains, you can see that all of the vectors are pointing in the same direction. The fact that all of the atoms in all of the domains are aligned means that all of the electrons' spins are also aligned, creating a billion tiny magnetic fields all going in the same direction. With all of the magnetic fields going in the same direction, a magnet will have an overall direction for its magnetic field, with one end being the north pole and the other end being the opposite, or south, pole.

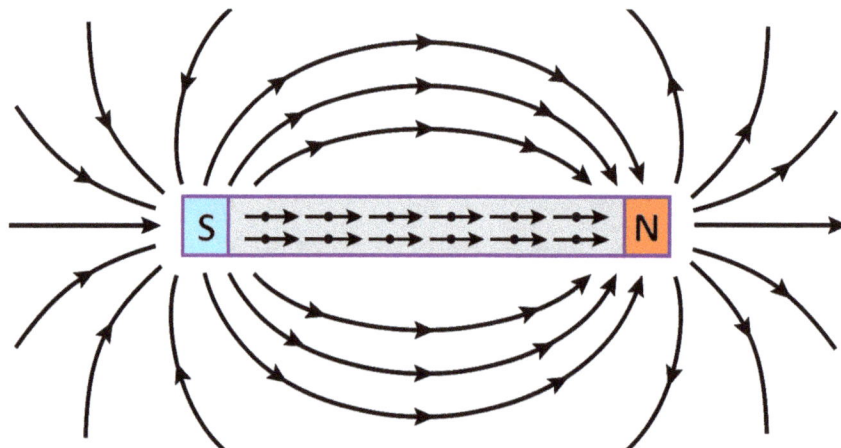

To find out what the magnetic field looks like, we can use the "left hand rule." If you hold up your left hand with your thumb pointing in the direction of electron flow, your fingers represent the direction of the magnetic field around that flow. If the electron flow is reversed, then the left hand is placed on the other side with the thumb pointing to the direction of the electric flow.

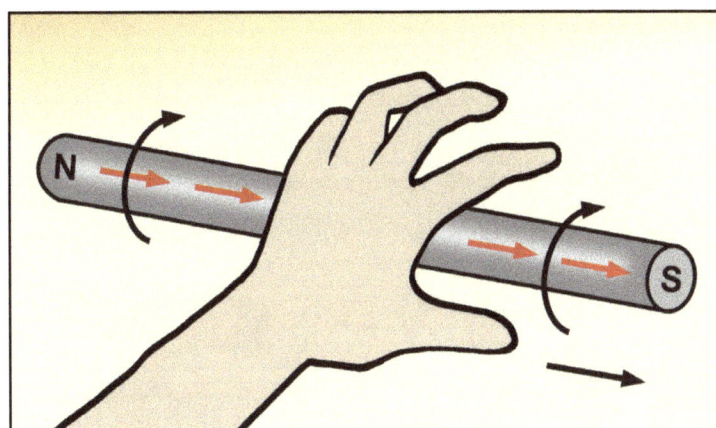

The same rules for attraction and repulsion apply to magnets as the ones that apply to electric charges—except the word "charge" is replaced with "pole:"

"Like" poles repel.

"Unlike" poles attract.

Magnets are also different from electric charges because magnetic poles cannot be separated. Recall that we can separate electric charges into positive and negative by using friction. We can rub a balloon across our hair and make the hair "charged." But we cannot separate magnetic charges. Friction won't work, and even if you cut the magnet in half, there will still be two poles, one on each end! You could keep cutting and cutting, but even the smallest piece you cut would still have two poles. You cannot separate the poles of a magnet. This is because the atoms themselves behave like little magnets.

12.3 Electromagnets and Electromagnetic Induction

As we just observed, any time an electric current flows through a wire it creates a magnetic field around the wire. A coil of wire with a current flowing through it behaves much like a bar magnet. So you can make a magnet, called an electromagnet, using just a battery and wire. If you take an iron rod, place it inside the coil of wire, and hook the wire to a battery, the iron rod will become an even stronger magnet.

NO, JOE. I SAID, "GET THE *ELECTRO*-MAGNET," NOT "GI-GAN-TIC-O MAGNET"!

Electromagnets can be very strong, so strong that they are used in junk yards to lift heavy cars. However, unlike regular magnets, electromagnets can be turned "on" and "off" by simply turning on or off the electric current in the wire. Once the electric current is gone, the magnetic fields have been turned off —they no longer exist. This can be very useful for junk yards!

If electric currents can induce metals to become magnetic, can magnetic objects induce wires to flow charges? Yes! This is called electromagnetic induction. When a magnet is moved up and down inside a coil of wire, it produces electric current. When a magnet is pushed through the coil, the electrons in the wire move back and forth as the magnetic fields push the electrons back and forth. This back-and-forth motion of the electrons produces an electric current that can be measured with a voltmeter. The more loops in the coil, the stronger the electric current that is produced. In fact, twice as many loops will create twice as much voltage. Also, the faster the magnet is pushed through the coils, the greater the voltage created.

Magnet is moved up and down through the wire coil

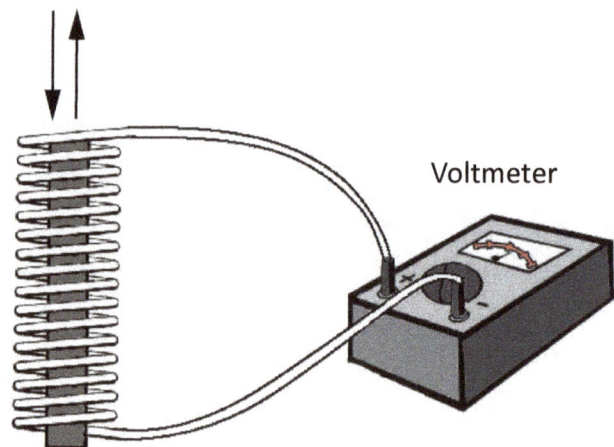

Voltmeter

Scientists and engineers can create powerful electrical generators utilizing electromagnetic induction. Early electrical generators were called dynamos. The word dynamo comes from the Greek word *dynamis* meaning power.

12.4 Earth's Magnetic Field and the Compass

Have you ever thought about hiking a trail, climbing a mountain, or floating down the rapids in a river? If you are planning an expedition, you might also think about how you can make your way through the wilderness and not get lost! One piece of equipment you could take with you to keep you going in the right direction is a compass.

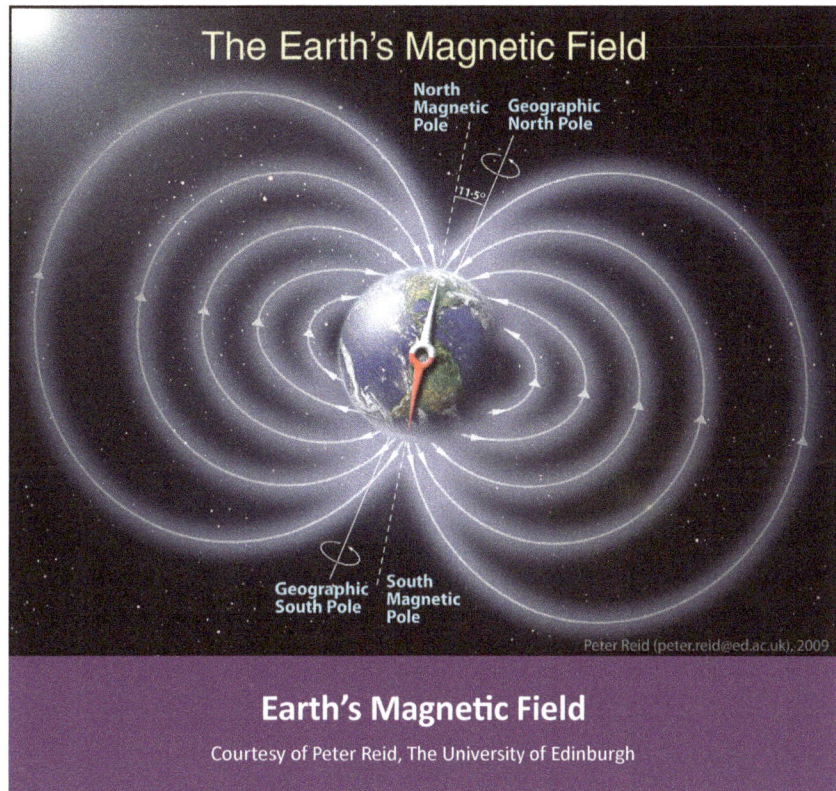

The Earth's Magnetic Field

North Magnetic Pole
Geographic North Pole
111.5°
Geographic South Pole
South Magnetic Pole

Peter Reid (peter.reid@ed.ac.uk), 2009

Earth's Magnetic Field
Courtesy of Peter Reid, The University of Edinburgh

A compass contains a small suspended magnetized needle that points toward north. If you learn how to use a compass correctly, you can always find your way while hiking a trail in the woods or climbing a mountain.

I THINK WE NEED TO GO THIS WAY.

YES. THAT'S RIGHT.

One way to look at Earth's magnetic field is to imagine Earth as one big magnet! Although the needle of a compass points northward, it does not actually point to true north. The Geographic North and South Poles are found at the ends of the imaginary axis that runs through the center of the Earth and around which

Earth rotates. True north is another name for the Geographic North Pole. Instead of aligning with the geographic poles, a compass aligns with the magnetic field of the Earth. If you imagine Earth as having a bar magnet running through the center in a north/south orientation, the compass would align with the north pole of the bar magnet. This is the Magnetic North Pole. The magnetic poles are close to the geographic poles but are in a slightly different location.

The magnetic field surrounding Earth is not quite constant like that of a bar magnet—it moves a little bit each year. Occasionally, it can move so much that the magnetic poles switch position, with the north magnetic pole becoming the south and vice versa. In fact, the poles have flipped many times in the past, with more than twenty reversals in the past five million years.

We don't know exactly why the Earth has a magnetic field. Although Earth's magnetic field acts similar to a big bar magnet, Earth's core is much too hot for the atoms to hold a magnetic arrangement there. Some scientists think electric currents deep inside the Earth loop around the molten iron in the core, creating a magnetic field. Other scientists think that electric currents are the result of heat rising from the core causing convection currents that create the magnetic field. No one really knows right now why the Earth has a magnetic field, and more research needs to be conducted.

12.5 Biomagnetism

Why can some pigeons be used to send messages? How do they find their way? Called homing or carrier pigeons, these birds have the ability to track back and forth between two locations. Pigeons have tiny magnetite crystals in their heads behind their beaks, and it is thought that they may use the Earth's magnetic field to sense their direction of flight. Biomagnetism is the use of magnetic fields by living organisms. By using Earth's magnetic field and their tiny inboard compass to orient themselves, biomagnetism may be how pigeons navigate.

Other organisms are also thought to use Earth's magnetic field as a way to navigate. A type of bacteria called magnetotactic bacteria create single domain magnetic crystals in small organelles called magnetosomes. The bacteria organize these crystals into long chains to form an internal compass. With this compass they can detect changes in Earth's magnetic field, equipping them with a sense of direction that helps them locate food.

Magnetobacteria were first observed by Salvatore Bellini, an Italian medical doctor, in 1963 when he was using a microscope to look for bacteria in water. He observed that a certain type of bacteria would always swim north and would then collect in a group at the north side of the water sample he was looking at. They would stop at the edge of the water sample and not go in a different direction.

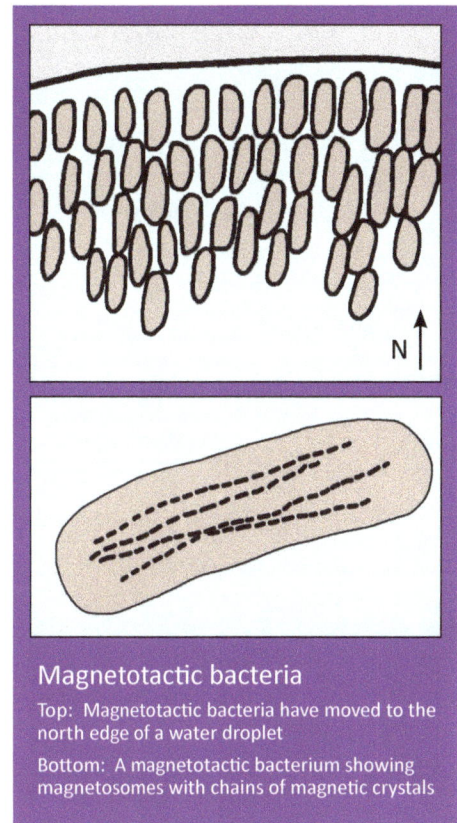

Magnetotactic bacteria

Top: Magnetotactic bacteria have moved to the north edge of a water droplet

Bottom: A magnetotactic bacterium showing magnetosomes with chains of magnetic crystals

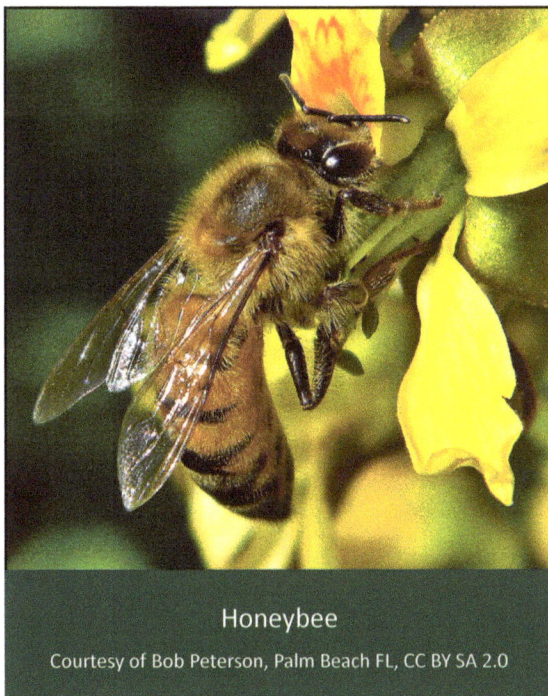

Honeybee

Courtesy of Bob Peterson, Palm Beach FL, CC BY SA 2.0

Honeybees also use Earth's magnetic field to navigate. A honeybee contains magnetite in the abdomen of its body, and although scientists do not know exactly how the honeybees use the magnetite, they show a remarkable ability to navigate using Earth's magnetic field.

12.6 Summary

- The word magnet comes from the Greek *magnetis lithos* meaning the stone of Magnesia.

- There are two types of magnetic objects—permanent magnetic objects and temporary magnetic objects.

- Spinning electrons create magnetic dipole fields.

- An unequal number of spinning electrons make some materials magnetic.

- Magnetic materials have magnetic domains.

- Electromagnetic induction is the generation of electric current by moving a magnet up and down inside a coil of wire.

- Earth has a magnetic field.

- A compass is a simple and very useful tool for outdoor activities.

12.7 Some Things to Think About

- Think about how you might have discovered a lodestone if you lived in ancient times. Write your story.

- Why are some materials magnetic and others are not?

- What is one way that magnets are different from electric charges?

- Explain how electromagnetic induction works.

- Do you think electromagnetic is a good term to use as a descriptor? Why or why not?

- If you were using a compass and a map to navigate a long hike in the wilderness, do you think the different positions of the Geographic North Pole and the Magnetic North Pole would have to be taken into consideration? Why or why not?

  Do you think the movement of the Magnetic North Pole would have to be taken into consideration? Why or why not?

- What other animals can you think of that might use biomagnetism? What might they use it for?

- Do you think human bodies might use biomagnetism? Why or why not?

Appendix: Math Solutions

Physics Math Problem 1   Ch. 7, p. 65

1. $\dfrac{10 \text{ m/sec.} - 25 \text{ m/sec.}}{|5 \text{ seconds} - 0 \text{ seconds}|} = \dfrac{-15 \text{ m/sec.}}{5 \text{ seconds}} = -3 \text{ m/second}^2$

2. $\dfrac{40 \text{ m/sec.} - 10 \text{ m/sec.}}{|10 \text{ seconds} - 0 \text{ seconds}|} = \dfrac{30 \text{ m/sec.}}{10 \text{ seconds}} = 3 \text{ m/second}^2$

Physics Math Problem 2   Ch. 7, p. 66

$v_i = 20 \text{ km per hour}$

$v_f = 6 \text{ km per hour}$

$t_i = 0$

$t_f = 0.10 \text{ hour (6 minutes)}$

What is the acceleration?

$$a = \frac{v_f - v_i}{|t_f - t_i|} = \frac{6 \text{ km/hr} - 20 \text{ km/hr}}{|0.10 \text{ hr} - 0 \text{ hr.}|} = \frac{-14 \text{ km/hr}}{0.10 \text{ hr}} = -140 \text{ km/hr}^2$$

Physics Math Problem 3   Ch. 8, p. 73

No. The center point doesn't travel any distance.

Physics Math Problem 4   Ch. 8, p. 75

1. Farther out

2. Both positions have the same rotational speed

3. It doesn't move. It has no speed.

Glossary-Index

[Pronunciation Key at end]

absolute value ('ab-sə-lüt 'val-yü) • a positive number, 55, 64

accelerate (ik-'se-lə-rāt) • in physics, to change in speed and direction, 24-25

acceleration (ik-se-lə-'rā-shən) • a change in velocity over a given time, 24-25, 53, 63-66, 69, 70, 137

accelerator • see particle accelerator

acid-base exchange reaction • the type of chemical reaction that occurs between an acid and a base—atoms of one molecule change places with atoms of another molecule to make new molecules, 79

algebra ('al-jə-brə) • a type of mathematics that uses symbols in arithmetic calculations, 13-14, 55, 56, 65

al-Khwarizmi (al-'kwär-əz-mē) • [circa 780 - circa 850 CE] Persian mathematician who is thought to have developed algebra, 13

ampere ('am-pir) • a unit of current (symbol for ampere is A), 119-122

angular speed • see speed, rotational

anode ('a-nōd) • the "high pressure" side of a battery (-), 112, 113, 118

antiproton (an-tē-'prō-tän) • a proton with a negative charge, 18, 19

Aristotle ('a-rə-stä-təl) • [384-322 BCE] Greek philosopher; studied motion and other physical laws, 46, 47

average speed • see speed, average

axiom ('ak-sē-əm) • a statement that is used as a starting point, premise, or postulate, 13

axis ('ak-səs) • a line around which something rotates, 9, 74, 128, 133

balanced forces • forces of equal strength acting on each other; shown by objects that are not moving or are moving at constant speed, 23-24

battery ('ba-tə-rē) • a cell that stores chemical energy; has the potential to create electrical pressure through chemical reactions, 11, 12, 32, 37, 90-91, 112-113, 122, 131

battery, electrochemical ('ba-tə-rē i-lek-trō-'ke-mə-kəl) • a cell that uses chemical reactions to produce energy, 90-91, 112

biomagnetism (bī-ō-'mag-nə-ti-zəm) • the use of magnetic fields by living organisms, 134-135

cadence meter ('kā-dəns 'mē-tər) • a device that measures how fast a bicyclist is rotating the pedals, 74

capacitor (kə-'pa-sə-tər) • an electrical component that stores electric charge, 16, 118

carbohydrate (kär-bō-'hī-drāt) • a molecule containing both carbon and water, 37, 89-90

carburetor ('kär-bə-rā-tər) • in an internal combustion engine, the part in which fuel and air are mixed, 87-88

cathode ('ka-thōd) • the "low pressure side" of a battery (+), 112, 113, 118

CERN (sern) • Conseil Européen pour la Recherche Nucléaire [European Council for Nuclear Research] a scientific organization that is exploring theories about the universe, 18-19

chemical energy • see energy, chemical

chemical potential energy • see energy, chemical potential

chemical stored energy • see energy, chemical potential

circuit ('sər-kət) • the path electricity takes through a wire, 12, 16, 112-113, 116, 118, 120

circuit ('sər-kət), **closed** • an electrical circuit through which electrons can flow, 113

circuit ('sər-kət), **open** • a broken electrical circuit through which electrons cannot flow, 113

circuit, parallel ('sər-kət, 'per-ə-lel) • an electrical circuit in which components are in parallel with their heads connected together and their tails connected to each other, 118

circuit, series ('sər-kət, 'sir-ēz) • an electrical circuit in which the components are are connected along a single path, 118

circular motion • see motion, circular

citric ('si-trik) **acid cycle** • the reaction in which ATP and NADH are made, 89, 90

closed circuit • see circuit, closed

combust (kəm-'bəst) • to burn, 86-88

combustion (kəm-'bəs-chən) • the process of burning, 86-88

compass ('kəm-pəs) • a device that has a magnetized needle that uses the Earth's magnetic field to tell direction, 127, 133, 134

compression stroke • see stroke, compression

computer (kəm-'pyü-tər) • an electronic device that stores, retrieves, and processes data, 7, 12, 16, 17-18, 125

conductor (kən-'dək-tər) • a material that allows electrons to flow, 104, 113-118

connecting rod • in an internal combustion engine, the part that attaches the piston to the crankshaft, 87-88

electrostatic induction (i-lek-trə-'sta-tik in-'dək-shən) • the effect of a charged object causing electric charges to move closer to it or father from it, 104-106

electrostatics (i-lek-trə-'sta-tiks) • [L., *stare,* "to stand"] the area of physics that studies electric charges that are not moving, 94-106

EMF • see electromotive force

energy ('e-nər-jē) • in physics, the ability to do work, 11-12, 22, 26-27, 37-43, 78, 80, 82, 87, 88, 89, 90

energy, chemical ('kem-i-kəl 'e-nər-jē) • energy that comes from chemical reactions, 11, 16, 37, 39, 78-91

energy, chemical potential ('e-nər-jē, ke-mi-kəl pə-'ten-shəl) • the energy stored in molecules; also called chemical stored energy, 31, 32, 37, 40-41, 85-91

energy, chemical stored • see energy, chemical potential

energy crisis • the disappearance of usable energy that is being converted into unusable forms, 39

energy, elastic potential ('e-nər-jē, i-'las-tik pə-'ten-shəl) • the power stored in a stretched object, such as a rubber band, or a compressed object, such as a spring; also known as strain potential energy, 31, 33

energy, electric potential ('e-nər-jē, i-'lek-trik pə-'ten-shəl) • the potential for an electric charge to do work; electrical pressure; voltage, 102-103, 112

energy, electrical • energy that comes from the flow of electrons; electricity, 11, 12, 16, 17, 31, 32, 37, 40, 41, 42, 43

energy, gravitational potential ('e-nər-jē, gra-və-'tā-shə-nəl pə-'ten-shəl) • a form of stored energy that requires gravity to convert it to another form of energy, 30, 32, 37, 38, 42, 102, 103, 110

energy, heat ('e-nər-jē, 'hēt) • the transfer of energy from one object to another; thermal energy, 11, 27, 39, 43

energy, kinetic ('e-nər-jē, ki-'ne-tik), [abbrev. **KE**] • [Gr., *kinetikos,* putting in motion] the energy of motion, 27, 32-34, 37, 38, 42

energy, light • energy from electromagnetic waves, 11, 16, 32, 37, 39, 43

energy, mechanical ('e-nər-jē, mi-'ka-ni-kəl) · energy having to do with the motion or position of an object, 11, 17, 32, 37, 42, 89, 90

energy, nonrenewable • a natural resource that once used up cannot be restored, 41

energy, nuclear potential ('e-nər-jē, 'nü-klē-ər pə-'ten-shəl) • the energy stored in an atom, 31

energy, potential ('e-nər-jē, pə-'ten-shəl) • capacity to do work; stored energy, 27, 29-34, 37, 40, 41, 85

energy, renewable ('e-nər-jē, ri-'nü-ə-bəl) • a source of energy that is built up again by natural processes — for example, energy coming from wind, water or sunlight, 42-43

energy, solar ('e-nər-jē, 'sō-lər) • energy from the Sun, 43

energy, stored • the capacity to do work; (see also potential energy), 29-30

energy, stored chemical • the energy stored in molecules; chemical potential energy, 85

energy, strain potential • elastic potential energy, 31, 33

energy, usable • energy that can be converted into a form that we can use, 39

engine, four-stroke • an internal combustion engine with four piston movements—intake, compression, power, and exhaust, 87-89

engine, internal combustion ('en-jən in-'tər-nəl kəm-'bəs-chən) • a type of engine in which gasoline burns within the engine to produce the energy to run it, 86-89

Euclid ('yü-kləd) • [circa 325-circa 265 BCE] Greek mathematician called the founder of geometry, 13

Euclidean geometry (yü-'kli-dē-ən jē-'ä-mə-trē) • a type of geometry developed by Euclid that is based on a set of axioms, 13

exhaust stroke • see stroke, exhaust

flowing water • a source of renewable energy, 42, 43

force • in physics, a power or energy that changes the position, shape, or speed of an object, 4, 8-9, 22-26, 29-30, 46-50, 78, 81, 97-99, 102-104, 110

force field • the area affected by a force, 99

force gauge ('fôrs 'gāj) • an instrument used to measure compression, tension, torque, and gravitational force, 8

force gauge, digital ('fôrs 'gāj, 'di-jə-təl) • an instrument that uses electronics to measure compression, tension, torque, and gravitational force, 8

force, gravitational • see gravitational force

force gauge, mechanical ('fôrs 'gāj, mi-'ka-ni-kəl) • an instrument that uses moving parts to measure compression, tension, torque, and gravitational force, 8

forces, balanced • see balanced forces

forces, unbalanced • see unbalanced forces

fossil fuels • substances found underground that formed from plants and/or animals subjected to extreme pressure over a very long period of time; oil, natural gas, coal, 39-41

Pronunciation Key

a	add	l	love	v	vase		
ā	race	m	move	w	way		
ä	palm	n	nice	y	yarn		
â(r)	air	ng	sing	z	zebra		
b	bat	o	odd	ə	a in above		
ch	check	ō	open		e in sicken		
d	dog	ô	jaw		i in possible		
e	end	oi	oil		o in melon		
ē	tree	oo	pool		u in circus		
f	fit	p	pit				
g	go	r	run				
h	hope	s	sea				
i	it	sh	sure				
ī	ice	t	take				
j	joy	u	up				
k	cool	ü	sue				

More REAL SCIENCE-4-KIDS Books
by Rebecca W. Keller, PhD

Building Blocks Series
yearlong study program — each Student Textbook has accompanying Laboratory Notebook, Teacher's Manual, Lesson Plan, Study Notebook, Quizzes, and Graphics Package

Exploring the Building Blocks of Science Book K (Activity Book)
Exploring the Building Blocks of Science Book 1
Exploring the Building Blocks of Science Book 2
Exploring the Building Blocks of Science Book 3
Exploring the Building Blocks of Science Book 4
Exploring the Building Blocks of Science Book 5
Exploring the Building Blocks of Science Book 6
Exploring the Building Blocks of Science Book 7
Exploring the Building Blocks of Science Book 8

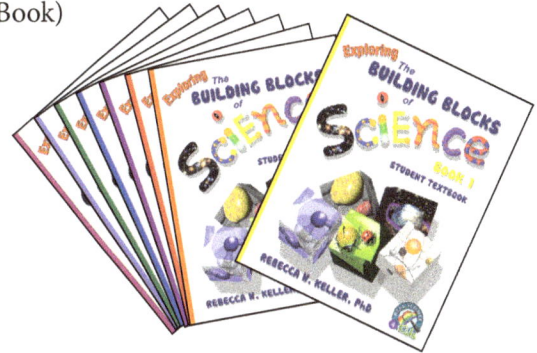

Focus Series
unit study program — each title has a Student Textbook with accompanying Laboratory Notebook, Teacher's Manual, Lesson Plan, Study Notebook, Quizzes, and Graphics Package

Focus On Elementary Chemistry
Focus On Elementary Biology
Focus On Elementary Physics
Focus On Elementary Geology
Focus On Elementary Astronomy

Focus On Middle School Chemistry
Focus On Middle School Biology
Focus On Middle School Physics
Focus On Middle School Geology
Focus On Middle School Astronomy

Focus On High School Chemistry

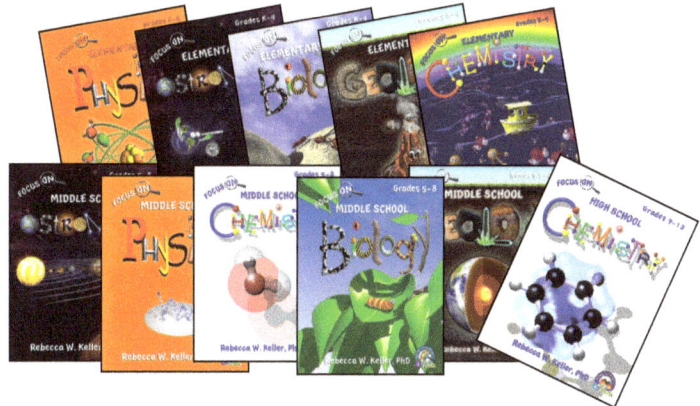

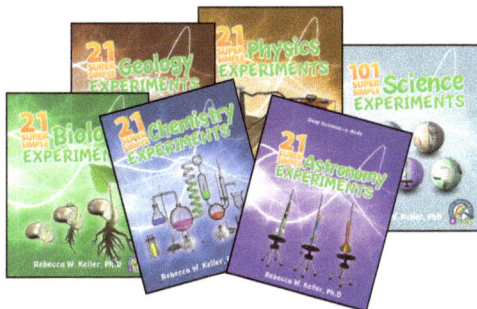

Super Simple Science Experiments

21 Super Simple Chemistry Experiments
21 Super Simple Biology Experiments
21 Super Simple Physics Experiments
21 Super Simple Geology Experiments
21 Super Simple Astronomy Experiments
101 Super Simple Science Experiments

Note: A few titles may still be in production.

Gravitas Publications Inc.
www.gravitaspublications.com
www.realscience4kids.com

www.ingramcontent.com/pod-product-compliance
Lightning Source LLC
Chambersburg PA
CBHW050240220326
41598CB00047B/7457